Photovoltaics for Professionals

Solar Electric Systems –
Marketing, Design and Installation

Falk Antony
Christian Dürschner
Karl-Heinz Remmers

Solarpraxis AG · Berlin · 2007 | Earthscan · London · 2007

Publisher:	Solarpraxis AG
	Zinnowitzer Str. 1
	10115 Berlin, Germany
	E-Mail: info@solarpraxis.de
	Internet: www.solarpraxis.de
	ISBN-13: 978-3-93459-543-9
in association with:	Earthscan
	8–12 Camden High Street
	London
	NW1 0JH
	UK
	Internet: www.earthscan.co.uk
	ISBN-13: 978-1-84407-461-7
Authors:	Falk Antony
	Christian Dürschner
	Karl-Heinz Remmers
	Solarpraxis AG
Translation and Adaptation:	Frank Jackson
Technical proofreading:	Doug Pratt
Graphics, layout, typesetting:	Solarpraxis AG
Printing:	Rondo Druck GmbH
Cover photo:	BP Solar
Copyright:	© 2007 by Solarpraxis AG

The texts and illustrations in "Photovoltaics for Professionals" have been prepared to the best knowledge of the authors. All photos and diagrams without citation have been created by Solarpraxis AG. Neither the authors nor Solarpraxis AG accept any responsability or liability for loss or damage to any person or property through using the material, instructions, methods or ideas contained herein, or acting or refraining to act as a result of such use.

Explanation of symbols

This symbol indicates a possible danger to persons, equipment or property, or possible financial loss.

This symbol indicates a tip or advice regarding design and installation practice, or marketing.

Foreword

This book is designed to be more than a technical manual on installing photovoltaic systems. Our priority was to provide designers and installers with practical specialist knowledge that would enable them to design and install high quality solar electric systems. However, we have also attempted to give an overview of the major photovoltaic market sectors and provide general guidance on marketing the technology and good business practice. While getting the technical side of installing solar electric systems right is essential, solar businesses also need to be commercially successful. The book has been structured with this in mind, as can be seen from the title of the opening chapter, *Marketing and promoting photovoltaics*. This is followed by *Solar cells, PV modules and the solar resource*. Subsequent chapters deal with designing, installing and maintaining systems. The bulk of the book deals with grid-tied systems, the largest market sector. A comprehensive chapter on stand-alone systems has also been included.

A knowledge of the environmental and energy security aspects of photovoltaics is important for solar business and these have also been discussed. Customers and potential customers are increasingly concerned about these issues and becoming increasing well-informed about them. Photovoltaic professionals involved in marketing the technology need to be able to discuss them knowledgeably.

The authors hope that their work will serve as a useful contribution to a change from a fossil fuel economy to one based on renewable energy.

Falk Antony   *Christian Dürschner*   *Karl-Heinz Remmers*

Contents

1 Marketing and Promoting Photovoltaics

1.1 Renewable energy

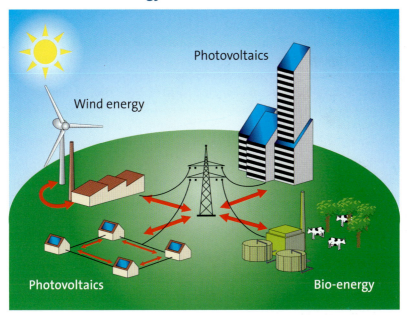

Figure 1.1: There is no shortage of renewable energy from the sun and the wind. Hydro energy, bio-energy and geothermal resources are also considerable

Our planet's renewable energy potential is effectively limitless when measured against humanity's needs. Solar energy, wind energy, hydro-electric, bio-energy (energy from plants) and geo-thermal energy are already being used on a large scale and in many different ways. Solar thermal technology uses solar energy directly to produce hot water. Photovoltaics (PV) produces electricity directly from sunlight. Plants harness solar energy and produce biofuels – wood, straw, vegetable oils for bio-diesel and other fuels. Wood itself is the oldest fuel. The use of water as an energy source goes back at least two thousand years and today hydro-electric power plants generate vast quantities of electricity. Most large potential hydro-electric sites are being utilized or developed but there is still enormous potential at medium and small sites. Wind turbines are the second largest generator of electricity from renewable resources. Total global installed wind power capacity (2004) is already 47.6 GW (Giga Watt = one billion watts). A 1.5 MW (MegaWatt = one million watts) wind turbine, with a rotor diameter of about 70 meters – can generate 76 GWh, gigawatt-hours, of electricity in 20 years; the amount of electricity a modern brown coal/lignite power plant would need to burn about 84,000 tonnes of fuel to produce.

1.1.1 Renewable energy around the world

Renewable energy presents business opportunities for manufacturers, distributors, system designers and installers. A whole range of technologies are already being produced, marketed, installed and creating jobs in Europe, Japan, North America and China. Every technology sector is experiencing significant growth.

The European Union (EU) has set itself a target of generating 12 % of its primary energy from renewables by the year 2012, with electricity generation from renewables to increase from 14 % in 1997 to about 22 % in 2010. Each member state has been allocated targets. How they achieve these targets is left up to the individual states. In May 2004, the European Commission announced that Germany, Spain, Denmark and Finland were on track to meet the 2012 target. 72.7 % of global wind generating capacity is in Europe.

Germany is the world leader in terms of renewable energy use and equipment manufacturing capacity. In 2004, wind energy overtook hydro energy as the main source of renewable energy electricity generation. The total installed wind generator capacity in Germany is now 16.6 GW. In 2004 wind energy produced 4.1 % of the country's electricity and hydroelectric plants produced 3.4 %. 6,300,000 m² of solar thermal collectors have been installed. 90 % of heat energy produced from renewables comes from bio-energy and bio-energy in the form of biomass and biogas generated 53,000 GWh of electricity in 2004. The German Renewable Energy Law, which offers guaranteed prices for PV-generated electricity has led to dramatic growth in this market sector – 360 MWp was installed in 2004 – making Germany the world's largest market for PV.

In the USA, about 6 % of total energy is produced from renewable sources, mainly hydro-electric and geothermal. There is over 7.2 GW of wind energy installed, 86 MW of PV and 52,000 m² of solar thermal collectors (2004). The USA is the world's third largest PV market. Japan is the second largest – 280 MWp was installed there in 2004. In the same year China installed 14,000,000 m² of solar thermal collectors.

One third of the world's population has no access to electricity at all – this is over 1,700 million people. Most of them live in rural areas and require relatively small amounts of electricity, too small to justify the expense of extending the grid. Small stand-alone PV systems are in reality the only practical way to supply many of these people with power. There is a large market for small stand-alone PV home systems in Asia, Africa and Latin America.

1.2 Climate change and dwindling fossil fuel reserves

Environmental problems associated with the burning of fossil fuel and the medium and long-term supply of fossil fuel itself, particularly oil and gas, make the move towards a global renewable energy economy both desirable and necessary. At the end of the 21st century, in only 400 years of industrial activity, humanity will have used up much of the fossil fuels which have been deposited in the earth's crust over the last 400 million years. Burning coal, oil and gas is releasing vast quantities of carbon dioxide into the atmosphere and changing the global climate. It is now clear that carbon dioxide in the atmosphere is increasing, and that the change in climate, both regionally and globally, is clearly measurable. The UN Intergovernmental Panel on Climate Change (IPCC) predicts a temperature increase of between 1.4 °C and 5.8 °C over the next 100 years, depending on whether industrialized societies and newly industrializing societies continue in a business-as-usual scenario or shift to a low carbon economy. The shifting of climate zones and the increasing frequency of extreme weather events such as floods, storms and droughts associated with climate change will severely damage the natural environments on which millions of people are dependant. The burning of fossil fuel also produces a range of other pollutants which impact on human health and the environment: benzene, soot, nitrogen oxides, hydro-carbons, carbon monoxide, sulfur oxides and ammonia. Acid rain destroys forests and lakes, and oil spills regularly cause massive environmental damage. A change in the way we use and produce energy is necessary to preserve the eco-spheres on which human life is dependant. Only a radical shift from fossil fuels to a low carbon economy will achieve this. The use of renewable energy will reduce carbon dioxide and other greenhouse gas emissions and help avoid a possible environmental catastrophe.

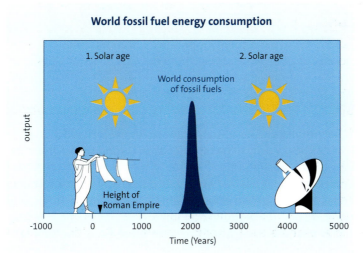

Figure 1.2: In a period of 400 years a large part of the fossil fuels which have been deposited in the earth over a period of 400 million years will have been used up

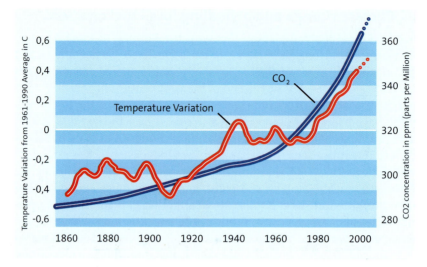

Figure 1.3: The increase of carbon dioxide in the atmosphere and the increase in temperature have been measurable for many years

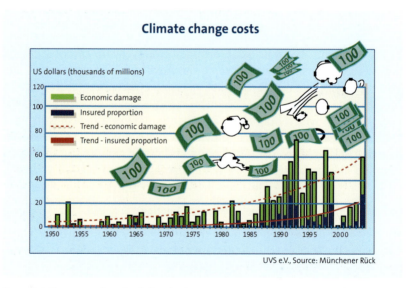

Figure 1.4: The economic costs of climate change

The other major problem is the actual supply of fossil fuel itself. Currently about 80 % of the energy used in the industrialized world comes from fossil fuel in the form of coal, natural gas and oil. It is now generally accepted that the reserves of oil and gas will be largely depleted within a few decades. Demand is increasing, particularly from the new industrializing and rapidly growing economies of China and India. Many of the reserves are also in politically unstable or relatively inaccessible regions. The supply of oil and gas could become critical in the coming years and lead to shortages. This is reflected in steadily rising prices.

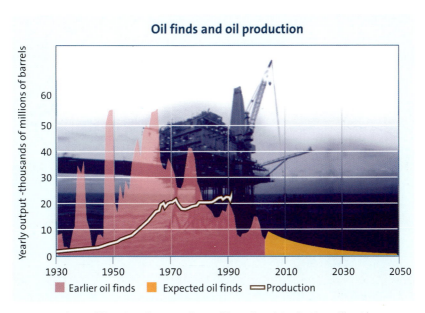

Oil finds and oil production

Figure 1.5: The world's major oil reserves have all been found. Production will not keep up with demand

1.3 Overview of renewable energy, solar energy and the solar resource

Most renewable energies are in effect solar energy – directly or indirectly. The wind is the result of the heating effect of the sun on the earth's atmosphere, as is hydropower. Plants (biomass) need sunlight to grow. Geothermal energy is an exception, it uses the heat of the earth's core. Although hydro turbines, biomass power stations and wind turbines utilize solar energy indirectly, the term *solar energy* is usually used to refer to devices which utilize the energy of the sun directly to produce both heat and electricity. These technologies are broadly categorized into two categories: *active solar* and *passive solar*.

Passive solar is differentiated from active solar in that it does not have any moving parts or electrical components. It usually refers to architectural and constructional features which reduce the overall energy needs of buildings by keeping them warm in winter and cool in summer. Passive solar architecture has a tradition of several thousand years. In ancient Greece buildings were constructed so that during the cold winters the low sun contributed to the heating of the building and during the summer the extended roof provided shade. Small windows (or none) on the north side of a building and large windows on the south side are typical passive solar features in northern latitudes. Passive solar technology is very site-specific – a passive solar house in Alaska will look very different from one in Florida.

Figure 1.6: Design for a house in Central Europe with several passive solar features, such as conservatories and small windows on the north side. The central heating system is solar thermal assisted (Source: http://www.elco.net)

Active solar systems are categorized into solar thermal, which usually produce hot water, but can also be used for cooling and producing steam which drives electricity-generating turbines – and photovoltaics (PV), which produces electricity directly from solar radiation.

Figure 1.7: Active solar technology: PV (left) producing electricity, and (right) a solar thermal system producing heat energy

Figure 1.8: Solar thermal power station in California. The parabolic mirrors concentrate the sun's rays on the pipe to produce steam to generate electricity (Source: Solar Millennium AG, Erlangen, D)

Solar hot water heating is the most common solar thermal technology. It produces hot water in homes, hotels and hospitals. It is also used to heat swimming pools even in sunny regions, thus extending their period of use. And it can also be used to produce hot water for industrial purposes. Solar water heating is particularly efficient at converting solar energy into heat energy. This gives it a relatively short energy pay-back period – 1 to 2 years. In many situations the hot water is usually used at the end of the day when it is available and it is easily combined with other water heating technologies such as electricity, gas and oil.

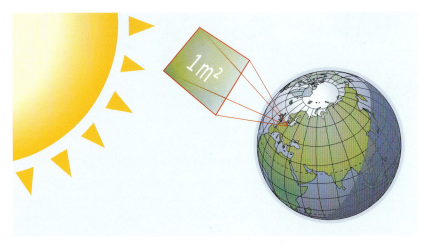

Figure 1.9: Solar is the most democratic of energy sources, falling fairly evenly on all the world's surfaces. A full midday sun averages about 1,000 watts per square meter

The amount of sunlight falling on the earth contains far more energy than we can ever use – 10,000 times more than the whole of humanity's current consumption. And with a life expectancy of billions of years, the sun is a source of energy which will effectively never run out. Even in northern latitudes, in regions not generally thought of as sunny, solar energy can be harnessed effectively. To supply current global electricity needs using PV, the land area needed would be only 1.5% of the European landmass, about 145.000 square kilometers (380 km by 380 km) or 56,000 square miles (237 miles by 237 miles), an area about the size of the US state of Iowa. And that is at solar radiation levels found in the northern latitudes. It would be even less at the equator.

PV module surface area needed to supply today's global electricity needs

380 km

380 km

Figure 1.10: A PV array covering 1.5% of the European land mass would be enough to supply current global energy demand (based on global electricity requirement of 17,300 TWh – tera-watthours, one trillion watthours (2005) and a 145,000 km² PV array in central Europe with an efficiency of 12% generating 120 kWh/m²/year)

The general arguments for solar energy can be summed up as follows:

• the distribution of solar energy on the surface of the planet – in full sun it can reach about a maximum of 1,000 watts of energy per square meter (W/m²) – is fairly evenly spread in comparison to the availability of other sources of energy
• solar energy is accessible to everyone after the initial investment, running costs are extremely low

- the global distribution of solar energy coincides largely with regions of highest human settlement
- solar energy is reasonably predictable, and the yearly yield is fairly constant – on the continental level, at any given time during the day, the sun is usually shining somewhere and solar electricity can be distributed from a region where the sun is shining to one where it is overcast (the same principle applies to wind power)
- the energy pay-back time (energy amortization) of modern solar electric and solar thermal technology is much less than the life expectancy of the equipment, which is more than 25 years; this makes these technologies into serious producers of energy; even at the current state of technological development, they produce far more energy than has gone into their manufacture
- the use of solar energy is not associated with any environmental risks – no oil spills, nuclear accidents and other man-made environmental catastrophes, not to mention climate change
- solar energy is good for international relations, it helps avoid military conflicts over oil, can reduce poverty and inequality and it is not a target for terrorist or military attacks.

1.4 Photovoltaics (PV)

Photovoltaics (PV) is the active solar technology which produces electricity from solar radiation using solar cells encapsulated in panels called *PV modules*.

1.4.1 PV system types

There are two broad categories of PV systems: *grid-tied systems* which are connected to the public electricity grid, and *stand-alone systems* which are not.

Grid-tied PV is essentially a technology used in urban, suburban and light industrial zones of the developed world. It needs a stable grid infrastructure. In terms of PV module sales this is the largest market sector.

Stand-alone PV is generally more rural and found where there is no grid. Major market sectors are telecom, remote homes and buildings of any kind, holiday cottages, national parks, recreational vehicles, small solar home systems (SHSs) in rural areas of the developing world. However, there is also a significant market for stand-alone PV in urban and suburban areas in the developing world – for street furniture such as solar powered bus shelters, parking meters, traffic signs etc. In fact, PV can be used to provide electricity practically anywhere on the planet, from the Arctic to the tropics.

1.4.2 PV's unique selling points

PV is a technology with many advantages and unique selling points. It is worth summarizing them:

- PV technology is mature, robust and reliable, has no moving parts and requires minimum maintenance
- no fuel or fuel supply chain needed
- PV systems are relatively easy and quick to install – particularly grid-tied systems
- the components used in PV systems have proved themselves during long use, they are weather resistant, UV resistant and can withstand extremes of temperature – in space exploration PV powers space station life support systems
- it is modular, systems can be any size – from a pocket calculator to a sports stadium
- stand-alone PV can supply power practically anywhere on the planet
- PV systems reduce emissions of the greenhouse gas carbon dioxide
- PV reduces pollution generally
- PV systems help save scarce resources
- PV generates more energy than it consumes – the energy consumed in the production of the modules is recouped within between 2 to 7 years, depending on location and system type
- PV modules are recyclable; various technologies exist which can recycle modules after at the end of their lives or if they have been damaged; the cells, the glass, the aluminum frames can be either reused or recycled
- PV promotes energy awareness – home owners with grid-tied systems become interested in low energy appliances; with stand-alone systems low energy appliances are essential
- PV systems add to the value of the building on which they are installed
- PV is a popular technology and is a good advertisement for renewable energy generally
- PV is a rapidly expanding market nearly everywhere, one in which different types of businesses can find a niche.

Figure 1.11: PV has been powering life-support systems on space stations for years (Source: PHOTO ESA / D. Ducros)

1.4.3 Building integrated photovoltaics (BIPV)

Building integrated photovoltaics (BIPV) could revolutionize building technology in the same way that concrete did over a century ago. There is already an abundance of examples and numerous BIPV systems are already on the market. These include roofing systems for both sloping and flat roofs, facades, noise protection barriers and semi-transparent PV systems for use in halls and foyers. BIPV offers new opportunities for the building industry – aesthetically pleasing and constructionally sound roofs and facades which also generate electricity. PV facades installed on prestigious commercial buildings can replace expensive cladding.

Figure 1.12: Building integrated photovoltaics (BIPV) could revolutionize construction technology (Source: www.solarintegration.de)

1.4.4 Future potential

PV technology is already technically mature and reliable but there is still enormous scope for future development. The PV industry is basically a branch of the microelectronics industry and its products can be compared with other microelectronic products such as mobile phones, the LCD-television set and the laptop. These technologies were technically mature shortly after they appeared on the market. Mass production made them rapidly affordable and technical innovation improved their performance. The LCD-television is rapidly replacing the cathode ray tube which has been with us for 50 years, and mobile phone cameras could soon largely replace the digital camera – itself a product that has only recently come onto the market.

Figure 1.13: Just as in other sectors of the microelectronics industry, mass production offers the possibility of lower PV prices and improved performance

Worldwide, new products and processes are being developed. Significant cost reductions and improved performance can be expected to result from:
- improved efficiency of cells and modules
- advances in thin film silicon technology
- further development of coatings, such as copper-indium-diselenide (CIS)
- the use of new and more effective coatings
- improved productivity and the larger manufacturing facilities for crystalline silicon cells and modules
- improved performance and mass production of system components such as grid-tie inverters.

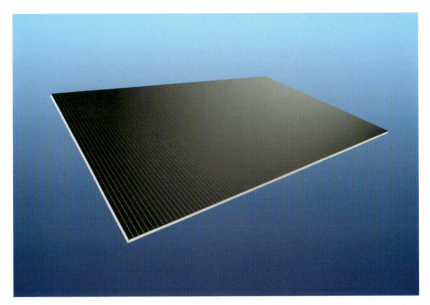

Figure 1.14: New products and processes are being developed worldwide, such as this cadmium-telluride (CdTe) module ...

Figure 1.15: ... and new production processes for cells

1.4.5 PV in non-sunny regions

The advantages of using solar energy in climatic zones which have a fairly regular distribution of sunshine throughout the year are obvious, but, despite subjective impressions, the use of solar energy in northern latitudes, where most of the sunshine is in summer and the days in winter are short and less sunny, also makes sense. In grid-tied PV systems, the seasonal variation in sunshine is not that important. It is the reliability of the energy source from year to year that counts. In Germany, for example, meteorological records show that from year to year total solar radiation rarely varies more than ±10% from the average. Between summer and winter, the change from the yearly average is of course substantial, and this does need to be taken into consideration when designing stand-alone systems.

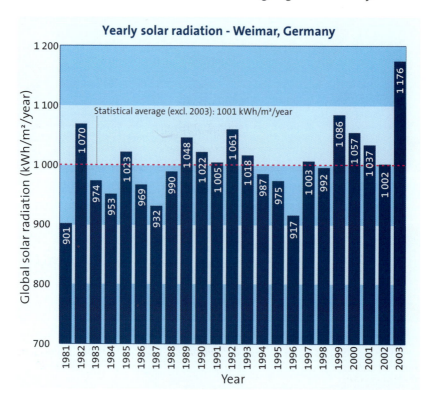

Figure 1.16: The sun is a very reliable source of energy for grid-tied PV systems. Records from Weimar, a northern latitude German city, show that solar radiation levels rarely vary more than 10% from the long-term annual average (Source: DWD)

There is more solar energy in the tropics and in hot deserts, where it is about 2,000 kilowatt-hours per square meter per year (kWh/m²/year). Typical figures for sunny locations are: San Francisco, USA – 1,700 kWh/m²/year; and in the southern latitudes: Sydney, Australia – 1,700 kWh/m²/year and Cape Town, South Africa – 1,900 kWh/m²/year. However, further north, there is also a significant solar resource: London, England – 1,000 kWh/m²/year; New York City, USA – 1,450 kWh/m²/year; Berlin, Germany – 1,000 kWh/m²/year; (average figures for solar radiation on the horizontal plane – PV modules are tilted towards the sun, so, in fact, it would be effectively more). And with increased efficiencies, rising conventional fuel prices and the more efficient use of energy generally, the attractiveness of PV will increase in the years to come.

Figure 1.17: In places like northern California and southern Spain about 1,600 kilowatt-hours of solar energy falls on every square meter of land surface annually – that corresponds to the amount of energy that is found in a barrel of oil. The barrel of oil bought and sold on the world market contains 159 liters of oil

There are very good reasons why the industrialized countries of the north should generate their own solar electricity:
- the sun is a reliable source of energy
- in many places it is more reliable and predictable than wind
- PV technology functions just as well in the northern latitudes as it does near the tropics; the photovoltaic effect actually works better in a cooler environment than it does in in hot desert conditions; and in addition, the arrays are cleaned regularly by the rain and are less likely to get covered in dust
- the industrialized countries possess the know-how and have the resources to invest in PV
- installing PV (and solar thermal) helps achieve energy security and reduces dependency on other countries, often ones in politically unstable regions

- installing PV promotes research and development into a technology which is increasingly being recognized as a technology of the future
- solar energy is generally popular – right now there is no other energy source about whose desirability there is such a wide consensus.

1.4.6 The economic viability of PV

The price of conventional fossil fuel energy is rising but the cost trend for PV is downwards. Reducing the cost of PV modules and associated system components depends on developing mass production techniques and facilities. A steadily growing and stable market is required. This is happening. The current global market for PV is over 1,200 MWp (2004) and it is growing at a rate of about 25 % a year. Studies by the Fraunhofer Institute in Freiburg, Germany, have indicated that in Germany the cost of electricity produced by PV could match the retail price that domestic consumers pay for electricity by about 2020. This could happen earlier in sunnier countries like Spain. The PV industry, from raw material to finished product to the actual installation of systems, creates wealth and high quality jobs and there are no hidden environmental costs. It also reduces dependence on imported oil and gas. (See also 2.4.12 *The cost of PV modules*, 3.11 *Preparing cost estimates and quotes*, and 6.1.3 *Future prospects for stand-alone PV systems.*)

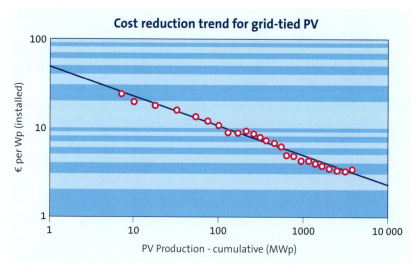

Figure 1.18: As more modules are produced and sold, the cost per installed kWp falls (Source: Fraunhofer Institute)

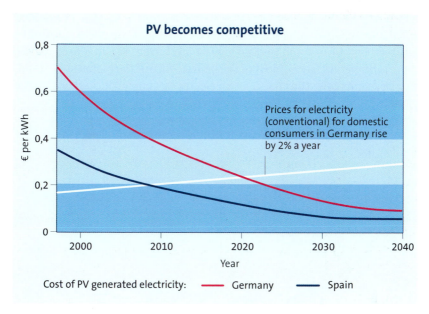

PV becomes competitive

Prices for electricity (conventional) for domestic consumers in Germany rise by 2% a year

Cost of PV generated electricity: —— Germany —— Spain

Figure 1.19: The cost of electricity produced by PV in Germany could match the retail price that domestic consumers pay for electricity in the foreseeable future (circa 2020). This could happen earlier in sunnier countries like Spain (Source: www.bsi-solar.de)

1.4.7 What energy generation might look like in the future

Up until recently mains electricity was produced exclusively by large central power stations and delivered to the consumer via high-voltage power lines. The growth of renewable energy technologies and combined heat and power plants (CHP) has changed this. Decentralized electrical generation technologies are becoming more common. These include CHP plants and fuel cells in building basements, or, in the case of PV, on the roof. Small decentralized generators have significantly lower distribution losses and some, like CHP plants can be brought on line very quickly at times of peak demand. They also reduce demand on the grid and offer greater security of supply. The grid of the future will look very different from today's. A combination of renewable energy technologies (solar energy, bio-energy, wind energy, hydro, energy from the waves, ocean currents and tides, and geothermal) could eventually provide 100% of our electricity needs. PV systems can easily be integrated into grid management, and weather forecasts can predict their output quite accurately. When PV output is high in the middle of the day other power sources (e.g. biomass) will be taken off line or, if not, their energy stored (e.g. pumped water storage for hydro) for use during the night. Besides being environmentally friendly, PV also produces electricity at the time of greatest demand. Economics, environmental legislation and security of supply issues will eventually determine the role of each particular technology in this mix. Dynamic technical developments are taking place. As the market develops, these technologies will improve and become more affordable. It is difficult to predict which one will be making the largest contribution in 50 years' time. Year by year the costs of renewable energy technologies are decreasing while those of fossil fuels and nuclear power are increasing. In places, where there is no electric grid – most of the land surface of the planet – PV stand-alone systems offer the chance of an electricity supply for everyone on the planet.

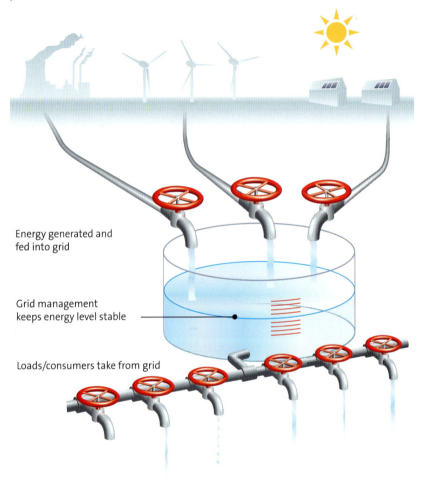

Energy generated and fed into grid

Grid management keeps energy level stable

Loads/consumers take from grid

Figure 1.20: The above graphic illustrates how the grid works by the use of a water reservoir analogy. All electricity producers – wind, PV, conventional power stations – fill a large reservoir with water (electricity). Grid management (i.e. putting generators on line when demand increases and taking them off line when it drops) ensures that the water in the reservoir is kept at a level which ensures that there is always sufficient water pressure (voltage) at the taps representing the individual consumers

Figure 1.21: What a renewable energy future might look like

1.5 Marketing PV

The world PV market has experienced enormous growth in the last few years, tripling in a period of only 3 years from 400 MWp (2002) to 1,200 MWp (2004). From the point of view of marketing and selling systems, there are two major categories: grid-tied systems and stand-alone systems. Besides the differences in technology – which are significant – customers interested in grid-tied systems have different requirements to those interested in stand-alone systems, the major difference being that the buyers of grid-tied systems already have electricity and customers for stand-alone systems usually do not, or at the most have a source of electricity (such as a diesel generator) which they are not satisfied with. The two technologies serve different sectors and different marketing approaches are required.

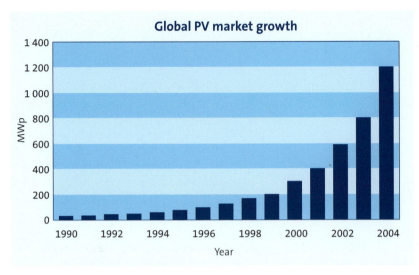

Figure 1.22: The global PV market: 1,200 MWp was installed in 2004. Annual growth is about 25 % (Source: Fraunhofer Institute)

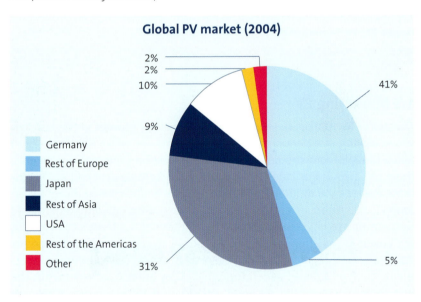

Figure 1.23: The global PV market is already worth € 6,300,000,000

When approaching the whole subject of marketing PV, it is worth bearing in mind that, for most customers, PV will be a new technology and most will know very little about it and PV systems are not cheap – a system represents a considerable investment. A professional approach is essential. Installers need not only to be able to design and install PV systems, but they also need to be knowledgeable about renewable energy and energy use in general, and national incentive schemes. On the other hand, system owners are proud of their PV systems and are the best advertisement for the industry.

1.5.1 Finding customers

The first task of a PV business is to find customers. How one goes about this will depend on many factors: the market sector – grid-tied or stand-alone, the state of development of the market the country or the part of the country one is in, whether one is an established business or a new-comer to the industry. For example, an established electrical business will already have customers, some of whom might be interested in installing a grid-tied PV system in a building which is being renovated, or rural customers who have been using a noisy fuel-guzzling generator, who might be interested in replacing it with a PV-diesel hybrid system. It is important to advertise. Magazines, local newspapers, free newspapers and local radio and TV can be used. A user-friendly web site is a usually a necessity these days. Most people will look at it before making direct contact with a solar business. Obtaining free publicity in local newspapers is also a possibility. Solar energy is new and what is new is news. A press release with photos of a new installation can form the basis of an article. Radio and TV stations are often interested in stories with an environmental angle.

Figure 1.24: Customers are usually proud of their solar systems and are the best advertisers (Source: www.willi-krauss.de)

1.5.2 Demonstration systems

A demonstration system is a very good way of showing what a PV system is and how it works. This is as true in the center of a city the size of London as it is in the middle of the African bush. A demonstration system can be one's own (in the house, office or workshop) or a friend's or a relative's. Having a system oneself demonstrates confidence in the technology. A demonstration system can also be the system of a customer.

Many customers are enthusiastic about the technology and will be happy to have it used in advertising and may even welcome site visits. A promotional folder containing photos and diagrams to show to potential customers is a good idea. The photos should be of both completed installations and work in progress, the diagrams should explain how systems works.

1.5.3 Initial consultancies with customers

First contact with a potential customer will usually be a telephone conversation, possibly followed up by a visit to the customer's home or business premises, if they are nearby. Most solar businesses receive lots of calls – only a minority of them will materialize into sales and installations, so it is important to ascertain if the query is promising before taking any further action. Site visits take up time, particularly if it the site is in a remote location. Essential information to gather at this stage is:

- why the customer is interested in a PV system – it may not be able to do what the customer wants it to do, e.g. heat a house
- and ascertain if there is a realistic budget available.

Some solar businesses send prepared forms to customers, asking them to provide essential information on the site and, particularly if it is a stand-alone system, what appliances need to be powered. Getting the customer to take a few photos of the potential installation site can be a great help. This is usually followed up by an initial discussion at one's office (where there might be a demonstration system) or on a site visit. (For the technical aspects of site visits see 3.6 *The site survey* for grid-tied systems and 6.4.1 *Feasibility assessment, site survey* for stand-alone systems).

The initial consultancy is an opportunity to establish confidence and demonstrate professionalism, which, of course, should be maintained throughout the course of the installation and during after-sales service. The following should help in this regard:

- showing the customer a folder containing photographs and diagrams of installations similar to the one the customer is interested in – many installers carry a *brag book* with pictures and details of systems they have done, and phone numbers of satisfied customers the prospective client can call
- providing information on any grants and financial incentives which may be available
- describing in clear language how PV works in general and how it will work in the type of installation the customer is interested in – most customers are unlikely to be familiar with the details of the technology
- outlining all the advantages of the technology – these may be financial, environmental or, in the case of a stand-alone system, simply not having to run a generator all the time
- taking all customer requirements and questions seriously
- being honest about the limitations of the system
- taking notes

- surveying the proposed site with a *Solar Pathfinder* or other device which will show if there are any seasonal shading problems.

Most customers will have a lot of questions. It is important that they are answered clearly and accurately without overloading the customer with technical details. Many of these questions will be technical and many will be specific to the codes, legal frameworks and funding. Many websites have FAQs (frequently asked questions) regarding PV. It is well worth studying these. Besides technical questions, customers will be interested in how local incentives work, the financial mechanics of selling electricity on the grid, so it is important to be well informed in this regard.

1.5.4 Contracts, quotes, estimates and insurance

Eventually, after several discussions and usually a site survey and some design work, an offer is made to the customer. This may be a more or less off-the-shelf system or one designed specifically for the customer. It is important to be aware that site surveys and designing systems take time. This is particularly true for stand-alone systems in remote locations – just getting there and back can take a day. Some companies charge a fee for site visits and design work, sometimes followed by a discount on the final price of the system.

Negotiation is part of the process and this may take time. The technology will most likely be very new to the customer and a lot of issues may need to be explained. It helps if the system description given to the customer is clear. The job may also include other work such as the installation of a solar water heating system. Keep this separate. Likewise, the whole system might, in fact be composed of several smaller systems – these should also be kept clearly separate from each other.

The more detailed the description of the proposed project, the less likelihood there is of problems arising in the future because of misunderstandings. If the customer is receiving a grant / tax rebate / preferential tariff, it is quite possible that the legislation covering these will also specify what details quotes and contracts should contain. In any case, customers will expect a system description to cover several pages and the more professional it looks, the better. It should be written in clear but exact language and describe the functions of the various system components and how they relate to the customer's requirements. Another advantage of a professional-looking and detailed system description in the initial offer to the customer is that the customer is quite likely to obtain several estimates and compare them with each other. The use of templates, which can be customized for each customer, will save work. Attaching copies of product specification sheets is also a good idea.

Contracts are legal documents and, for large systems, considerable sums of money may be involved. In the case of customers receiving government grants / tax relief / preferential tariffs for electricity sold onto the grid, many of the details will be specified by legislation.

Depending on circumstances, the contract may need to contain the following:
- the price of the system and all associated works
- a system description, with diagram
- description of main system components with reference to data sheets
- estimated energy output – this may need to be demonstrated by computer simulation
- grid connection issues, if it is a grid-tied system
- product and system warranties and guarantees
- details of module mounting structure
- array shading issues – masts or overhead cables may need to be removed, or trees trimmed.
- location of equipment such as inverters and batteries
- general layout / cable plan
- dates for beginning and end of work
- scaffolding requirements, other building and electrical works to be done by third parties
- payment arrangements agreed upon
- details of any building permits or structural surveys required or obtained
- lightning protection, if necessary
- details of installer status, e.g. *approved installer* status may be necessary
- details of maintenance agreements, if any
- insurance issues.

Figure 1.25: A 12 kWp PV roof being installed at the Centre for Alternative Technology in Wales, UK. The costs of ancillary work such as scaffolding and temporary structures necessary for working on roofs need to be included in estimates and contracts (Photo: Frank Jackson)

Grid-tied PV systems usually need very little maintenance. Stand-alone systems have batteries which need some maintenance if the system is not to develop problems. In many situations maintenance contracts make sense and will improve the efficiency of the system and prolong its life. If maintenance is not carried out, it may affect the warranties on the system components. A maintenance contract may include regular system monitoring. Regular monitoring and recording of system performance can be invaluable in identifying the cause of any problems which may develop. Some systems owners will be quite happy to do this themselves, other will not. Besides being an extra source of income, maintenance contracts help the installer keep in touch with customers – which might lead to more business – and give the opportunity to see the performance of systems over time.

Insurance is also an issue which may need to be addressed. PV systems are expensive and, though very robust, they can be damaged by lightning, a fire, a serious storm, a flood or vandalism. PV modules are valuable and can be stolen. An expensive PV array installed on or integrated into the roof of the building may not be covered by existing building insurance. Insurance can also be an issue in publicly-funded projects or where there is involvement of an incentive scheme. Many insurance companies will not have any experience in this field and may charge expensive premiums or refuse to offer coverage. However, this is an opportunity for the installer to find an insurance company that does offer coverage at reasonable rates and make putting the customer in touch with them part of the overall service. Installers will also need insurance themselves, such as public and product liability.

1.5.5 The market for grid-tied PV
Grid-tied PV is now the largest and fastest growing sector of the global PV market. Two of the major factors behind this growth are the German Energy Law (see 1.6 *Case study: Germany, a solar success story*) and the Japanese 70,000 solar roofs program. Currently, the most important markets are Germany, Japan and the USA. The installed capacity in these countries is currently: Germany 760 MWp (2004), Japan 1,100 MWp (2004), and the USA 270 MWp (2003).

Figure 1.26 –1.28: The market for grid-tied PV is mainly urban or suburban. *1* Office building facade in Munich, Germany, *2* Church in Barvaria, Germany, *3* A housing estate with grid-tied PV system in Japan (Sources: *1* www.architekten-pmp.de, *2* Peter Hasenbrink, *3* Hakushin Corporation)

There are many reasons why institutions and individuals are interested in having a grid-tied PV system installed on their building or home, and, from a marketing point of view, it is important to know what these are. These include:

- individual concern for the environment
- the desire of institutions and companies to demonstrate their green credentials
- a source of income from selling electricity onto the grid
- *offset* – i.e. customers want to supply some of their own electrical needs
- providing security of supply in connection with a back-up power system
- educational / demonstration
- adding to the value of a building.

For the prospective installer, it is essential to know why the customer wants the system and what budget is available. If these two questions can be clarified right at the beginning, it can save a lot of unnecessary work. For example, if a customer thinks he or she will make a lot of money from selling to the grid and this is not the case, it is unlikely that the sale will go through. And it takes time to design systems – spending a day or more designing a system which is well beyond the budget of the customer is not good business practice.

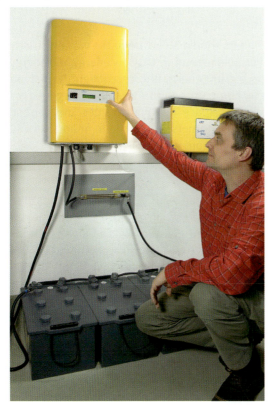

Figure 1.29: Grid-tied PV systems can be used as power back-up systems where grid outages are likely to last quite a while and power requirements are low, e.g. to keep a central heating pump working, or a credit card machine or internet connection. The system has a small battery bank (battery cover removed for clarity) but the array is so large in comparison that even on dull days there will be more than enough power to charge it. It has two inverters: an SMA grid-tied inverter (right) and an SMA stand-alone inverter-charger (left). For some customers this kind of system might be very attractive (Source: Wagner & Co. Solartechnik)

Net-metering, offset and low energy electrical appliances

Net-metering is the process of buying and selling electricity onto the grid. Some countries have laws covering net-metering, others do not. The crucial issue is what price is paid for electricity fed onto the grid in comparison to the price paid for electricity taken from the grid. It may be the same, it may be less or it may even be more. If it is higher, then the customer will naturally want to sell all the electricity generated by the PV system. If it is the same, it makes little difference whether the customers sells it all or uses it within the building. If it is less, the customer will want to use as much as possible within the building.

Using electricity from a PV array instead of purchasing from the grid is known as *offset*, i.e. a building's electricity is offset by that produced by the PV array. Indeed, in some systems so much of the electricity is used in the building that no net metering arrangements are made at all. So little is put onto the grid that it is not worth metering. What happens technically is that the meter slows down or stops when the PV array is producing enough electricity for the building's needs. Some older meters will actually run backwards, but doing this without authorization is usually illegal. The net-metering arrangements are crucial in working out the amortization of the system and will also effect the physical layout of the metering and physical connection to the grid (see also 4.6 *Connection to the grid and meter location*).

PV system installers and designers should be aware of the way electricity is used within a building and be able to recommend low energy and energy efficient options to customers, e.g. low energy fluorescent lighting instead of incandescent, flat screens for computers and TV sets, low energy fridges and avoiding using electricity for space heating. Besides the general environmental benefits of these, customers will also gain financially and the overall PV installation will make more sense.

1.5.6 The market for stand-alone PV

The market for stand-alone systems is also large, both in the industrialized and the developing world. Most of the surface of the planet is not serviced by an electricity grid. In the USA, 35 MWp of stand-alone systems have been installed, globally it is about 70 MWp, and accounts for 10 % of the global PV module market. Major suppliers estimate that about 600,000 stand-alone PV systems are installed each year (of varying sizes) and that the annual market growth rate is 10–15 %. In Africa and many other parts of the developing world, the market for stand-alone PV systems has been growing steadily over the last ten years.

Figure 1.30: In many areas of the developing world where people have no access to electricity at all, PV can provide an economic and secure power supply (Source: Energiebau Solarstromsysteme GmbH)

Since stand-alone PV systems can provide power anywhere, no short list of applications can be complete, but generally it would include:
- small solar home systems in the developing world
- schools, hospitals in the developing world
- medical and veterinary refrigeration
- solar home / holiday cottage systems in the developed world
- telecommunications
- solar water pumping
- street furniture – bus stops in rural areas, parking meters in cities
- solar lanterns and other portable appliances.

Stand-alone solar systems are also found in recreational vehicles, boats, remote weather stations and remote airports. Sometimes stand-alone PV systems, particularly the larger ones have an additional power source, usually a fossil-fueled generator or wind. These are known as PV-hybrid systems.

Figure 1.31: Stand-alone solar street lights are found at rural bus stops, parking areas and other off-grid locations

Marketing stand-alone PV systems differs from marketing grid-tied systems mainly in the fact that the potential customers do not usually have an electricity supply at all, or they have an electricity supply with which they are unhappy, such as a fuel-guzzling and noisy diesel generator.

Some of the advantages of stand-alone PV systems which can be pointed out to customers are:
- when properly sized, they offer a secure supply of electricity
- they can be practically any size, from a single security light to a large home or several buildings
- they can easily power most home and office electrical equipment
- silent and fuel free operation
- low maintenance
- can be installed almost anywhere
- they can be cheaper than obtainig a grid connection.

Figure 1.32: Solar boat in Berlin, Germany. Solar boats are silent and there are no accidental oil spills – ideal for environmentally-sensitive lakes, rivers and canals (Source: Solarpolis)

Figure 1.33: PV- powered road vehicle – the Honda Dream Machine (Source: Solarpolis)

Site visits are not always possible and in remote locations take up a lot of traveling time. Having a pre-printed form which customers can use to list power and energy requirement details can be very useful (see 7.5 *Site survey form – stand-alone PV systems*).

Figure 1.34: BP Solar portable solar lantern, Maasailand, Tanzania (Photo: Frank Jackson)

Off-the-shelf smaller stand-alone PV systems offer many advantages. They do not need to be individually designed anew for each customer and can be provided with an instruction booklet for DIY installation. The use of low energy and energy efficient electrical appliances is essential in stand-alone systems and installers and suppliers can offer these as part of the package, in particular low energy 12 and 24 VDC lights.

Figure 1.35: PV-powered tsunami early warning system buoys in the Indian Ocean (Source: A Rudloff, GFZ Potsdam, Germany)

1.5.7 Opportunities for marketing other renewable energy technologies

PV system installers are inevitably asked all sorts of questions about other renewable energies. This should be seen as a marketing opportunity. The most obvious technology to market alongside PV is solar thermal for water heating. Customers often need both electricity and hot water. Solar thermal and PV go well together, and many suppliers offer both in a combined package. Solar water heating systems for swimming pools are also attractive; even in sunny regions they extend the period of use of the pool at both the beginning and end of the season. With grid-tied customers there is usually room on the roof for a well-designed solar thermal system which can save on increasingly expensive oil or gas and significantly reduce household CO_2 emissions. For the stand-alone customer, offering a solar thermal system is a convenient and fuel free way of producing hot water.

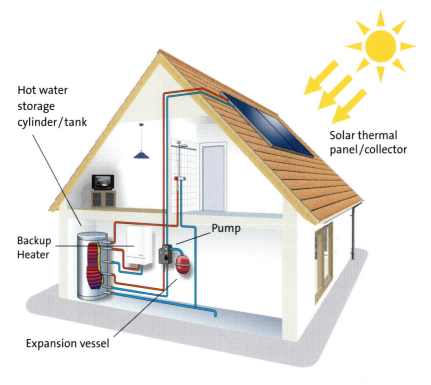

Hot water storage cylinder/tank

Solar thermal panel/collector

Pump

Backup Heater

Expansion vessel

Figure 1.36: Domestic solar water heating system

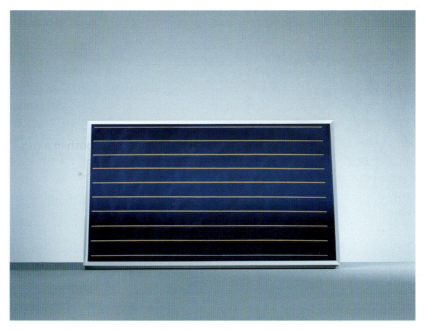

Figure 1.37: Solar water heating – flat plate panel (Source: Vaillant)

Figure 1.38: Solar water heating – evacuated tubes (Photo: SCHOTT AG/Stefan Kiefer)

Some customers will also be interested in space heating technologies such as wood and wood pellet heating systems and heat pumps. If a PV system is being installed on a building that is being renovated, this is an excellent time to discuss these possibilities. Grants from local or national government may be available. Customers in remote locations who are having a stand-alone PV system installed will also very often be interested in heating the building – they will have no access to piped gas and there may be a local supply of wood. There are many excellent wood-fuel heating systems on the market.

Figure 1.39: A growing number of manufacturers and suppliers are offering combined solar thermal and PV systems with PV modules and solar collectors. Some of theses systems can be integrated into the roof (Source: Schüco International KG)

1.6 Case study: Germany, a solar success story

In 2005, 450 MWp of PV was installed in Germany, the most PV generating capacity ever installed in a single year anywhere. That is about 3.6 million m² of PV modules with a total value of approximately € 1,800,000,000. At the time of writing, total installed capacity in Germany was over 1,000 MWp (end of 2005). That is over 100,000 PV grid-tied systems, with 20,000 new systems being added each year.

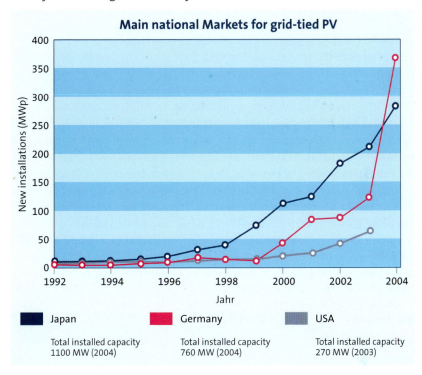

Figure 1.40: Growth in the German PV industry was spectacular after the Renewable Energy Law came into force. The other two main markets are Japan and the USA (Source: Fraunhofer Institute)

The growth in the number of PV installations and of the PV industry in Germany was a result of the country's innovative Renewable Energy Law. The law set up a legal and financial framework designed to increase the amount of electricity produced from renewable sources to 12 % by 2010 and to 20 % by 2020. At the time of writing (2005) it is 11 %. All renewable energy technologies that produce electricity sold onto the grid are covered by the law, not just PV. Previous attempts to kickstart the PV industry in Germany had not been very effective. Periodic initiatives which gave grants to system purchasers (up to 70 % of system costs in some cases) resulted in short-term intensive bushfire-like activity but did not lead to steady market growth. When each initiative had run its course, the market tended to stagnate again.

Under the new legal and financial framework created by the law:
- utilities were obliged to give preference to producers of electricity generated from renewable resources, including PV and facilitate the connection of generating plants to their networks
- prices were guaranteed for a fixed period of time – different technologies had different price structures
- the scheme was financed by a small surcharge on all electricity bills – about € 0.0042 per kWh to cover wind, hydro and biomass and € 0.0005 per kWh for the PV component.

The owner of a PV system is guaranteed a very attractive price for all electricity fed onto the grid during the first 20 years of the system's operation. This price is set at the time of system commissioning – at the time of writing € 0.55 approx. per kWh, depending on the type and size of the installation. It is more for systems mounted on building facades, which are more expensive to install, than free-standing structures or arrays on flat roofs. Obviously the economics of this would prove unsustainable if the number of PV installations increased beyond a certain point and the same price was guaranteed to everyone indefinitely, so the price is progressively reduced by about 5 % (6.5 % for free-standing arrays) each year for newcomers to the scheme . So, for example, a system coming on line in 2005 might be guaranteed a price of € 0.55 per kWh until 2025, while systems coming on line in 2006 would be guaranteed a price of € 0.52 per kWh until 2026. The overall effect of this was to make installing a PV system an attractive investment equivalent to a secure investment with a rate of interest of approximately 4–6 %.

Figure 1.41: As a result of an effective and sustainable renewable policy, tens of thousands of German homes are now solar powered (Source: SunTechnics GmbH)

The initiative also ensures that PV systems are optimally designed, installed and maintained – the better they work the more income they will generate for the owner. It has also led to the installation of larger systems with associated economies of scale – 5 kWp systems on single family homes are not untypical. There have also been many medium-sized systems installed – in the 20 kWp to 100 kWp range. These have very often been the result of co-operative enterprises organized by private citizens. In these solar investment clubs, each participant owns a part of a larger installation, installed on a roof other than their own, also providing an income for the owner of the roof. Each individually-owned part of the larger roof has its own meter and the owner of it receives the income from it – it is essentially a private PV installation but integrated into a technically similar larger one, which keeps initial costs down. There has also been considerable investment in even larger systems. These, in the MWp range, have been financed by solar investment companies for whom PV offers a lucrative and secure income.

For the development of a PV industry, the German Renewable Energy Law is a good example of how public policy can effectively and affordably support the growth of a PV industry. It provides a sustainable and growing market which enables forward planning, investment in research, development and mass production facilities, all of which are essential to bring down the costs of modules and inverters. Spain has now adopted a similar PV initiative which included a tariff of € 0.39 per kWh for systems up to 100 kW and a 25-year payment guarantee. The Spanish target is to install 400 MW by 2010. The California Solar Initiative, which was passed in January 2006, aims to install 300 MW per year in California alone for the next 10 years. It offers gradually diminishing installation rebates, but no performance payments.

Figure 1.42: Germany has seen considerable investment in larger systems – in the MWp range. These free-standing systems have been installed on brown field sites – old mines, industrial areas etc. (Photo: Dürschner)

2 Solar Cells, PV Modules and the Solar Resource

2.1 Basic PV concepts and terminology

The basic element of a PV system is the solar cell or photovoltaic cell. Exposed to light, they produce DC electricity. These are assembled and incorporated into photovoltaic or PV modules which are also commonly called *solar modules* or *solar panels* – but the later term can be misleading as it is also used to refer to the solar thermal flat plate collectors used in solar water heating systems. PV uses the light from the sun not the heat. All the PV modules connected together in a system are referred to as the photovoltaic or PV array, or *solar array* and sometimes as a *solar generator*. The whole system is known as a photovoltaic or PV system, which can either be connected to the electricity network (grid-tied) or not (stand-alone). The terms *solar system* and *solar electric system* are also used. Larger systems are sometimes referred to as *solar power stations*.

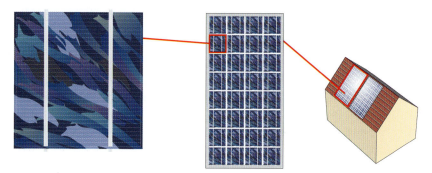

Figure 2.1: Solar cell – solar module – solar array.

The most common type of solar cell is the crystalline silicon cell. The first stage in their production is the manufacture of extremely thin wafers of silicon (Si) with a thickness of between 0.2 and 0.3 mm from high purity silicon. These wafers then undergo further processes to be made into solar cells with two semiconductor layers. When light falls on the cells, the electrical charges in the semiconductor layers separate and produce a DC voltage. The voltage produced is specific to the cell material; for silicon this is approximately 0.5–0.6 VDC. This is normally too low to be of use. In a PV module the cells are connected in series to produce higher voltages. The upper limit on the number of cells and thus the voltage of a module is determined mainly by the practicalities of handling modules. They must be light enough and small enough to enable them to be installed on roofs, sometimes under difficult conditions. To produce the even higher voltages required in grid-tied PV systems, modules are in turn connected in series. A group of modules in series is called a *string*. Larger installations will consist of several strings of modules.

2.1.1 Cell efficiency, module efficiency and system efficiency

It is important to differentiate between cell efficiencies, module efficiencies and system efficiencies. Commercially mass-produced solar cells can achieve an efficiency of 20 %, that is, they will convert 20 % of the sunlight falling on the surface area of the cell into electrical energy. Module efficiency is less because not all of the module surface area is covered by cells. There are spaces between the cells and the module frame also takes up space. And there are slight transmission losses getting the electricity from the cell surfaces to the module output cables. System efficiency refers to the overall efficiency of the installation. System losses can be caused by the use of different types of modules in an array, shading of the PV array, voltage drop in cables, inverter losses and resistances in connections, switches, breakers and fuses, and in batteries in stand-alone systems. People in the industry also differentiate between *field efficiency* and *laboratory efficiency*, i.e. the efficiency which can be expected under real conditions as opposed to what which can be achieved under ideal conditions in a laboratory.

2.2 PV module technology

Cells can be manufactured from about a dozen different types of materials. By far the most significant is crystalline silicon. At least 800 different PV modules produced by over 100 manufacturers are available on the international market. The three main types of commercially available cells are:
- monocrystalline silicon cells
- polycrystalline silicon cells
- thin film – amorphous silicon (α-Si) and materials such as copper-indium-diselenide (CIS), cadmium-telluride (CdTe) and gallium-arsenide (GaAs).

Mono- and polycrystalline silicon modules are the most common. They account for about 93 % of all modules sold globally and are used in both small and large systems. Thin film modules account for the remainder; amorphous silicon (α-Si) 4.2 %; copper-indium-diselenide CIS) 0.7 % and cadmium-telluride (CdTe) 1 %.

Monocrystalline modules are between 1.5 % and 2 % more efficient than polycrystalline, but the latter are slightly cheaper to produce and their share of the market is growing. However, from a practical point of view, neither is particularly preferable to the other. In terms of efficiency, cells can be categorized as follows:

1. Monocrystalline silicon
2. Polycrystalline silicon
3. Copper-indium-diselenide (CIS)
4. Cadmium-telluride (CdTe) thin film
5. Amorphous silicon (α-Si) thin film

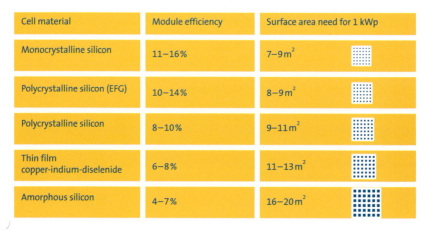

Cell material	Module efficiency	Surface area need for 1 kWp	
Monocrystalline silicon	11–16%	7–9 m^2	
Polycrystalline silicon (EFG)	10–14%	8–9 m^2	
Polycrystalline silicon	8–10%	9–11 m^2	
Thin film copper-indium-diselenide	6–8%	11–13 m^2	
Amorphous silicon	4–7%	16–20 m^2	

Figure 2.2: Cells made from different materials have different efficiencies. PV array surface area depends on the type of cell used. EFG (Edge-Defined Film-Fed Growth Technique) is a more efficient production process for polycrystalline cells

2.2.1 Monocrystalline silicon

To produce monocrystalline silicon a crystal of silicon is grown from highly pure molten silicon. This single crystal cylindrical ingot is cut into thin slices between 0.2 and 0.3 mm thick. These shiny silvery wafers are the basis of the monocrystalline solar cell. The use of these round cells in a module would give rise to non-productive spaces between the cells, so the edges are cut off to give a (usually) hexagonal shape so more can be fitted onto the module. Mass produced monocrystalline cells have an efficiency of between 13 % and 17 %, and they are the most efficient type of cell. However, they do require more energy and time to produce than polycrystalline cells. The most common cell size is 5 inches (125 mm), but 6 inches (152 mm) are now coming onto the market.

2.2.2 Polycrystalline silicon

Polycrystalline silicon (sometimes called *multi-crystalline*) is also produced from a molten and highly pure molten silicon, but using a casting process. The silicon is heated to a high temperature and cooled under controlled conditions in a mold. As the molten silicon sets, an irregular poly- or multi-crystal forms. This is visible in the shimmering fish-scale like surface appearance of the wafers. The square silicon block is then cut into 0.3 millimeter slices. The typical blue appearance is due to the application of an anti-reflective layer. The thickness of this layer determines the color. Blue has the best optical qualities. It reflects the least and absorbs the most light. Further chemical processes and the fixing of the conducting grid and electrical contacts complete the process. Mass-produced polycrystalline cell modules have an efficiency of between 11 % and 15 %. The square cells usually measure 5 inches (125 mm) or 6 inches (152 mm) across.

2.2.3 Thin film cells

Thin film technologies, such as copper-indium-diselenide (CIS), and cadmium-telluride (CdTe), despite their lower efficiencies (and thus larger array surface areas), are a promising alternative to silicon. They are much more resistant to the effect of shade and high temperatures, and offer the promise of much lower production costs.

2.2.4 Amorphous silicon

Amorphous silicon (α-Si) is non-crystalline silicon. Cells made from this material are found in pocket calculators and watches. Efficiency lies between 6 % and 8 %. The layer of semi-conductor material is only 0.5–2.0 μm thick. This means that considerably less raw material is necessary in their production than for crystalline silicon cells. The film of amorphous silicon is deposited as a gas on a surface such as glass or aluminum or plastic. (It is often simply called *thin film*). Further chemical processes, the fixing of a conducting grid and electrical contacts follow. Tandem (multi-junction) amorphous thin film cells with each layer sensitive to different wavelengths of the light spectrum are also available. These have slightly higher efficiencies.

*Figure 2.3: Top left – monocrystalline, top right – a polycrystalline cell (Source: Siemens)
below – a cadmium-telluride (CdTe) thin film module (Source: Antec Solar GmbH).*

Figure 2.4: Roll of Uni-Solar multi-junction thin film (Photo courtesy of United Solar Ovonic)

Type of cell	Construction	Cell Efficiency *	Module Efficiency	Current stage of development
Monocrystalline silicon	Uniform crystalline structure – single crystal	24%	13–17%	Industrial production
Polycrystalline (multi-crystalline) silicon	Multi-crystalline structure – different crystals visible	18%	11–15%	Industrial production
Amorphous silicon	Atoms irregularly arranged. Thin film technology	11–12%	5–8%	Industrial production
Gallium-arsenide	Crystalline cells	25%	**	Produced exclusively for special applications (e.g. space craft)
Gallium-arsenide, gallium-antimony & others	Tandem (multi-junction) cells, different layers sensitive to different light wavelengths	25–31%	**	Research and development stage
Copper-indium-diselenide	Thin film, various deposition methods	18%	10–12%	Industrial production
Cadmium-telluride & others	Thin film technology	17%	9–10%	Ready to go into production
Organic solar cells	Electrochemical principle based	5–8%	**	Research and development stage – not commercially available

Cell efficiency is based on laboratory samples, and is invariably higher than module efficiency. From the practical point of view of evaluating systems, the module efficiency should be used.

*** Not available in module form.*

2.3 Solar cells

2.3.1 Crystalline solar cells – what they are made of and how they work
Most solar cells are made of crystalline silicon of a very high purity, of the type used for the manufacture of semiconductors in the electronics industry. The original raw material is quartz sand (SiO_2), a material which is readily available. The raw silicon obtained from this must be made chemically pure before it can be used.

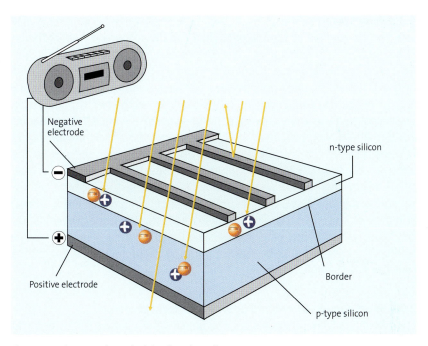

Figure 2.5: The operating principle of a solar cell

During manufacture the cells are *doped* with small quantities of other atoms, usually boron and phosphorous, to create two layers of silicon with different electrical characteristics: a positive (p) layer of p-type silicon and a negative (n) layer of n-type silicon. A conducting grid is fixed above and below to enable electron flow, and they are coated with an anti-reflective layer. At the border between these two layers (known as the p-n junction) an electrical field is created. On exposure to light the charges in the electrical field separate. This gives rise to a voltage of about 5.0 VDC between the electrical contacts on the cell. The value of this voltage difference is largely independent of the intensity of the light falling on the cell.

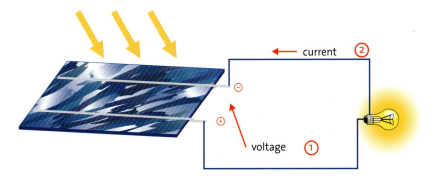

Figure 2.6: When exposed to light a voltage difference (1) occurs between the side of the cell exposed to the light and its underside. If a circuit is completed, electrical current will flow (2)

2.3.2 Peak power

The electrical output of a cell is proportional to the amount of solar radiation falling on it. It is at its highest in conditions of direct radiation (strong sunlight and clear skies). In conditions of diffuse solar radiation (cloudy weather), output is lower. The maximum power a cell (or a module) is likely to produce is described as its peak power (Wp). This is defined as the electrical output in watts achieved under Standard Test Conditions (STC) – 1000 W/m² solar insolation at cell temperature of 25 °C, and an Air Mass of 1.5. Air Mass (AM) is a measure of the thickness of the atmosphere which influences the spectral composition of the sunlight which reaches the earth's surface. Wp is also known as the *peak wattage* or *peak watts* (Wp) of the cell (or module/or array). A cell with a surface area of 100 cm² (10 cm × 10 cm) and an efficiency of 15 % will produce 1.5 watts (W) under these conditions. Output is directly proportional to cell surface area. A cell with twice the surface area and with the same operating characteristics will produce twice as much electrical energy.

2.3.4 Short circuit current and open circuit voltage

The principal electrical properties of a photovoltaic cell are its current and voltage. The current produced by the cell depends on the quantity of light falling on the cell and the size of the surface area of the cell and on the voltage at which it is operating. The cell output voltage is relatively independent of the level of solar radiation the cell is exposed to. That is specific to the cell material; for silicon this is approximately. 0.5–0.6 VDC. Higher voltages are achieved by connecting cells in series.

The relationship between current and voltage in a cell is described by its I-V curve. When the cell is short circuited, it produces its maximum current – its *short circuit current*, denoted by I_{SC}. When it is in open circuit, it produces what is called its *open circuit voltage*, denoted by V_{OC}, Under conditions of stable solar radiation, the current produced by the cell is determined by the operating voltage. And the maximum power point (MPP) of the cell is the point at which current and voltage produce the maximum power. The current at MPP is denoted by I_{MPP} and the voltage at MPP is denoted by V_{MPP}. The same principle applies to PV modules.

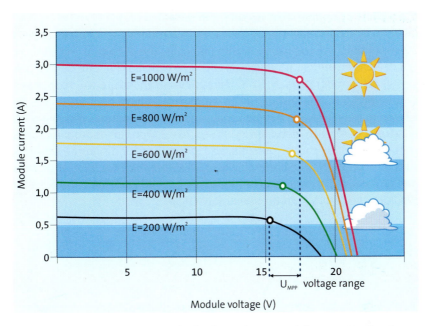

Figure 2.7: The I-V curve of a PV module. The electrical output and the current produced are directly proportional to the level of solar radiation. The voltage changes very little.

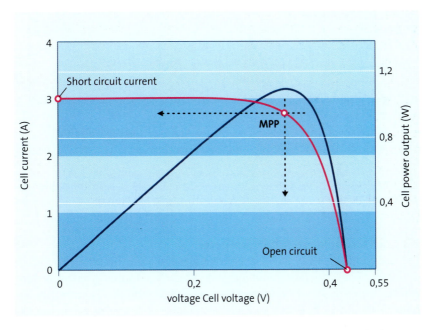

Figure 2.8: The I-V curve of a crystalline silicon solar cell. The open circuit voltage (V_{oc}) is about 0.5V. At the maximum power point (MPP) of the curve, the voltage is about 80% of the open circuit voltage (V_{oc}) and the current is about 95% of the short circuit current (I_{sc}).

2.3.5 Effect of temperature on solar cell output

Cell efficiency decreases with increases in temperature. Crystalline cells are more sensitive to heat than thin film cells. The output of a crystalline cell decreases approximately 0.5 % with every increase of one degree Celsius in cell temperature. A 30 °C increase in temperature will reduce cell output by 15 %. The output of an amorphous silicon cell decreases by approximately 0.2 % per degree Celsius increase. In summer, module temperatures can reach between 40 °C to 70 °C. For this reason modules should be kept as cool as possible, and in very hot conditions amorphous silicon modules may be preferable (see 3.4 *PV module mounting structures and systems*, 4.3.4 *Installing the modules – general guidelines* and 6.3.1 *PV modules and arrays in stand-alone systems*.

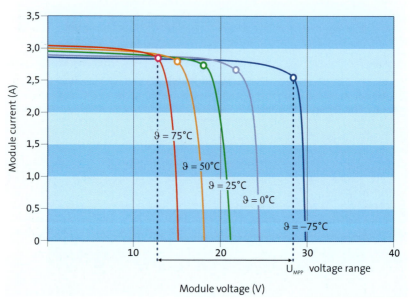

Figure 2.9: *The effect of temperature on the I-V curve of crystalline silicon PV modules. The output drops by about 0.5 % with every increase of one degree Celsius, but it increases at colder temperatures*

2.3.6 Cell technology – future prospects

The economic viability of PV depends essentially on the cell efficiency. A doubling of cell efficiency would result in nearly a doubling of the system yields. Cell efficiency in commercially available modules is currently between 5 % and 17 %. However, cells with higher efficiencies are being developed. Techniques such as structuring the surface of crystalline cells in order to minimize reflection, and trenching the conducting grid to reduce shade, have increased efficiency. Siemens / Shell has been using a coating that puts little pyramids over the cell to reduce reflection for 2–3 years now, and BP produces modules with laser-cut grids. Higher performing gallium-arsenide (GaAs) and gallium-antimony (GaSb) cells, composed of several layers – tandem or multi-junction cells – have already achieved efficiencies of 31%. More economic production methods are also being intensively researched. Cells based on promising materials such as copper-indium-diselenide (CIS) and cadmium-telluride (CdTe) are already being produced in small quantities. Research is also being carried out into the development of organic solar cells which use an electro-chemical process similar to photosynthesis. At the time of writing, it is impossible to predict which thin film and amorphous technologies will eventually become commercially significant.

2.4 PV modules

A PV module is composed of interconnected photovoltaic cells encapsulated between a weather-proof covering (usually glass) and back plate (usually a plastic laminate). It will also have one or more protective bypass diodes. The output terminals, either in a junction box or in the form of output cables, will be on the back. Most have frames. Those without frames are called *laminates*. In some, the back plate is also glass, which gives a higher fire rating, but almost doubles the weight.

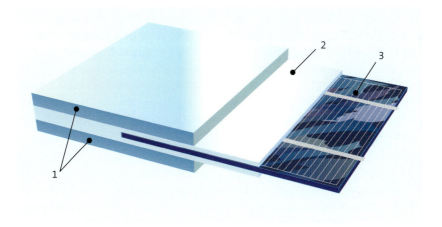

Figure 2.10: 1 Both front and back plate, both glass, 2 Encapsulation in ethyl-vinyl-acetate (EVA), 3 Crystalline solar cell.

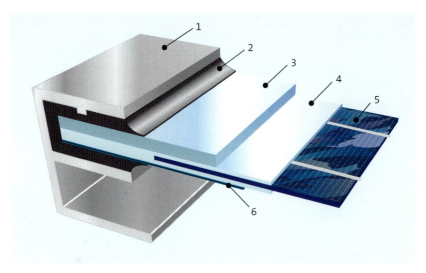

Figure 2.11: 1 Aluminum frame, 2 Seal, 3 Glass, 4 EVA, 5 Solar cell, 6 Tedlar sheet

The cells in the module are connected together in a configuration designed to deliver a useful voltage and current at the output terminals. Cells connected in series, also called *strings,* increase the voltage output; while cells connected in parallel increase the current. Each string of cells is usually protected by a bypass diode. These prevent damage through overheating occurring to the module if a cell is shaded or defective – the so-called *hot spot* effect – and limits the effect of the subsequent drop in output in the module concerned and other modules in the string. The maximum number of cells that can be incorporated into a module is determined by the desired size and weight. Modules need to be practicable to handle as they are sometimes installed on roofs which are rather inaccessible, or by only one or two persons.

The front of the module is usually covered with hardened or tempered plate glass. The back consists either of another plate of glass or a sheet of Tedlar. In order to ensure that they survive prolonged exposure to the elements – during an operational life of at least 25 years – the cells are encapsulated in an airtight layer of ethyl-vinyl-acetate (EVA). Whether the module has a frame or not depends on the mounting structure onto or into which it will be fixed or incorporated. At the back of the module (or laminate) or attached to the frame, there is a junction box containing the output terminals – or, in order to enable easier connection of modules in strings, multi-contact weatherproof flying leads with double-insulated and polarized DC connectors. Modules are available in a range of sizes, starting at about 5Wp. Those used in grid-tied systems range from 80 Wp to 300 Wp.

Figure 2.12: Five different types of solar cells and their physical appearance in different light conditions. Above: in sunshine. Below: in cloudy conditions. From left to right: Polycrystalline silicon, monocrystalline silicon, copper-indium-diselenide (CIS) thin film, amorphous silicon (α-Si), cadmium-telluride (CdTe) thin film. The five arrays have a peak power of 1kWp (Photo: Ernst Schrimpff)

2.4.1 Choosing a module – main points

In principle, when selecting modules these days, there is not a lot that can go wrong. Quality is high and a comprehensive system of international product certification is in place. The main things to pay attention to are:

• the quality of the finished product (visual inspection)
• certification/standards – independent product assessments are also useful
• the manufacturer
• application – grid-tied or stand-alone
• 20 to 25 year warranty.

From the point of view of product quality, the main requirements are that:

• modules enable electrically sound performance of the installation
• have long working lives, are suitable for the environment in which they will be installed and suffer minimal degradation over time in terms of performance
• meet required technical specifications

Figure 2.13: A damaged solar cell. These days most manufacturers have years of experience and pictures like this belong to the past (Photo: Volker Quaschning)

2.4.2 Which module manufacturer?

When deciding which manufacturer to purchase from, the following should be considered:

- their experience/length of time in the industry
- endorsements from previous customers
- the quality of service offered, including ease of contact (telephone, office hours, local representative)
- warranties and guarantees
- how long replacement modules will be available if needed
- frequent name changes in the past – this calls for caution
- extent of insurance covering warranties and guarantees.

This last point is often overlooked. The insurance of warranties and guarantees – to ensure that they are honored in the event of a manufacturer going out of business at some time in the future – is an important issue. Before making a final decision regarding a manufacturer/supplier, one should examine the small print of the warranty and make sure that it is adequate. Ambiguous points and omissions should be queried.

2.4.3 Which type of solar cell?

Mono- and polycrystalline modules are installed on both family and multi-occupancy dwellings. Neither has a particular advantage over the other. The type chosen is very often simply the one most readily available. Monocrystalline has a slightly higher efficiency than polycrystalline, which means that the modules will require a slightly smaller surface area. Some architects and builders will also prefer the appearance of the monochrome gray or blue color of the cells for aesthetic reasons. Polycrystalline modules are slightly cheaper. And some architects and builders prefer the distinctive shimmering appearance of the blue cell.

There is nothing to be said against the use of amorphous or thin film or either, however they require substantially more roof area and experience with them is not as long. Their output is considerably less affected by high temperatures and shading. Some builders and architects will prefer their overall homogenous appearance.

 Customers are increasingly interested in the aesthetics of solar arrays. Different types of modules look better on different types of roofs and facades. Advise the customer on what is available and the differences between types. A module can even be temporarily placed on a roof to see what it looks like.

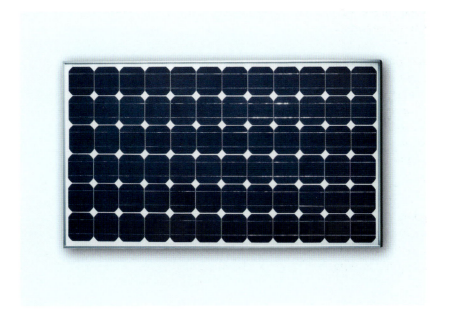

Figure 2.14: Monocrystalline modules have a higher efficiency and thus require slightly less surface area than polycrystalline modules (Source: Vaillant).

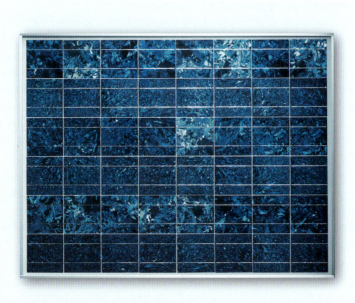

Figure 2.15: Some architects and builders prefer the distinctive shimmering appearance of polycrystalline modules (Source: Vaillant)

Figure 2.16: Multi-junction amorphous laminate on flat-roofed office building (Photo courtesy of United Solar Ovonic)

2.4.4 PV module electrical characteristics

Modules with high peak wattages are practicable, especially on large arrays. Fewer of them are required for a given array size, so the installation can be completed in a shorter period of time. Power tolerances are also important. ±5% is standard, but ±3% is not unusual these days. Modules whose output varies significantly on the minus side can seriously reduce the total output of the array due to mismatching.

In grid-tied systems module voltage and current ratings need to be such that the modules can be series-connected in strings of specified numbers which match inverter input requirements. High module maximum voltages (V_{MAX}) – the maximum overall voltage of a string of series-connected modules into which a module may be incorporated – may be required if a relatively high system voltage is chosen, as might be done to reduce losses in cables (see 3.2.1 *Grid-tied inverters and PV module configurations*). In stand-alone systems, the module voltages need to be suitable for charging the type of batteries in the system (see 6.3.1 *PV modules and arrays in stand-alone systems*).

2.4.5 Temperature tolerances, hail impact resistance, module weight and dimensions

A low temperature coefficient (the measure of decrease in output against increase in temperature) is particularly important if the modules or laminates are to be integrated into the roof structure or installed in any situation where there will be no airflow under the array. Resistance to water, abrasion, hail impact and other environmental factors is also essential. Module weight needs to be considered, particularly if the module is to be installed by one person and/or roof accessibility is problematic.

The external dimensions of modules are more important than is commonly assumed. Larger modules reduce the number of points at which they need to be fixed to a roof or structure and thus help reduce installation costs. However, they do reduce layout options. Finding the ideal module to fit a particular roof can take up a considerable amount of system design time. Modules also need to fit around existing structures on the roof, such as skylights, dormer windows, plumbing vents, and possibly solar water heaters. And access to the roof might be restricted – through a narrow skylight for example.

Figure 2.17: If both PV modules and solar thermal collectors are to be installed on the same roof, an attempt should be made to chose sizes which fit reasonably harmoniously together (Source: Roto)

If a non-standard or customized size or type of module is to be used – for example, if modules are being integrated into a customized mounting structure on a roof or facade – one needs to ensure that replacements, if required, will be available in the future. In some cases, replacements will need to be purchased by the customer and kept in storage.

2.4.6 PV module frames and inter-module electrical connections

Framed modules offer the advantage of a wider range of mounting possibilities. The frames themselves can be used as support and fixing points. This is particularly useful on older roofs with irregular angles and planes or where the rafters are irregularly spaced. Frames also offer additional protection against the elements.

PV laminates (modules without frames) are sometimes preferred for aesthetic reasons. They are integrated into roofs and installed on the facades of buildings. Aesthetically pleasing and homogenous roof surfaces can be achieved, and the sometimes unacceptable reflection of sunlight from the aluminum frames of modules avoided. The rain will also clean the array more effectively – strips of accumulated dirt above the lower edges of module frames are less likely. Snow will also slip from the smooth surface with greater ease. Both these points are particularly important on roofs with shallow inclinations.

Modules and laminates with multi-contact weather-proof flying leads with double-insulated and polarized (reverse-polarity-protected) DC connectors (a plug-socket arrangement) may be slightly more expensive than modules with junction boxes containing screw terminals. They take less time to install and there is less possibility of connecting them up incorrectly when they are being incorporated into long strings. Another advantage of this type of inter-module connection is that the contacts do not loosen over time – this reduces the possibility of faults developing and the associated cost of rectifying them. Modules with junction boxes containing screw terminals are usually preferable in smaller stand-alone systems as they make it easier to connect up the modules in the parallel and series-parallel configurations often required for battery-based systems (see 6.3.1 *PV modules and arrays in stand-alone systems*).

2.4.7 PV module/system compatibility

Modules need to fit the mounting structure and be compatible with the other electrical components in the system. In grid-tied systems the electrical specifications of the modules should be compatible with those of the inverter, and those of the inverter with the modules. In stand-alone systems they need to be compatible with the charge controller. It is worth considering, where possible, purchasing complete systems from a single supplier, but one should make sure that all the components supplied are compatible.

2.4.8 PV module specifications and their importance

The European Standard EN 50380 specifies exactly what technical specifications a data sheet for a PV module should contain. This information is needed for system design and estimating the yearly or monthly energy yield. The specifications will also usually be required in applications for grants and financing. In reality, some module data sheets do not give all the information specified in EN 50380, and North American specification requirements are less complete. (Temperature coefficients are not listed.) Manufacturers and suppliers should be asked to confirm them in writing.

EN 50380 requires that values for the following module characteristics be given on module data sheets:
- Wp or W   nominal power output or watts peak, with ±power rating tolerance
- V_{MPP} voltage at the maximum power point
- I_{MPP} current at the maximum power point
- V_{oc} open circuit voltage
- I_{sc} short circuit current
- temperature coefficients for power output, current and voltage.

All these values should be given for Standard Test Conditions (STC) – 1000 W/m² solar insolation at a module operating temperature of 25 ºC, and an Air Mass of 1.5. If possible the values given should be for the particular module, not general values for the module make or type.

The temperature co-efficient is required because, on a sunny day, modules temperatures can rise to 70 ºC causing a reduction in module output. On the other hand, output will increase at low temperatures. In fact, on a sunny winter day, the peak output can be higher than on a sunny summer day. This temperature effect needs to be taken into consideration in system design and in estimating the yield of an array. The presence of by-pass diodes in the module is also important. These enable current to by-pass shaded cells, reducing the effect of shading (see 3.10.3 *Solutions to shading*).

Module data sheets should also say what certification the module has, in addition to describing the cell material, the frame material and the type of glass, giving the module dimensions (length, width and depth) and weight. While these details are not important for estimating output, they are necessary for the design and planning of the physical installation. The designer needs to know:

• the number modules that can be fitted onto the roof
• if the load-bearing capacity of the roof is sufficient for the array
• what module mounting structure/fixing options might be suitable.

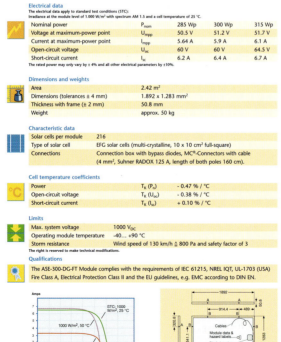

Electrical data
The electrical data apply to standard test conditions (STC):
Irradiance at the module level of 1.000 W/m² with spectrum AM 1.5 and a cell temperature of 25 °C.

		285 Wp	300 Wp	315 Wp
Nominal power	P_{nom}	285 Wp	300 Wp	315 Wp
Voltage at maximum-power point	U_{mpp}	50.5 V	51.2 V	51.7 V
Current at maximum-power point	I_{mpp}	5.64 A	5.9 A	6.1 A
Open-circuit voltage	U_{oc}	60 V	60 V	64.5 V
Short-circuit current	I_{sc}	6.2 A	6.4 A	6.7 A

The rated power may only vary by ± 4% and all other electrical parameters by ±10%.

Dimensions and weights

Area	2.42 m²
Dimensions (tolerances ± 4 mm)	1.892 x 1.283 mm²
Thickness with frame (± 2 mm)	50.8 mm
Weight	approx. 50 kg

Characteristic data

Solar cells per module	216
Type of solar cell	EFG solar cells (multi-crystalline, 10 x 10 cm² full-square)
Connections	Connection box with bypass diodes, MC®-Connectors with cable (4 mm², Suhner RADOX 125 A, length of both poles 160 cm).

Cell temperature coefficients

Power	$T_K (P_n)$	- 0.47 % / °C
Open-circuit voltage	$T_K (U_{oc})$	- 0.38 % / °C
Short-circuit current	$T_K (I_{sc})$	+ 0.10 % / °C

Limits

Max. system voltage	1000 V_{DC}
Operating module temperature	-40... +90 °C
Storm resistance	Wind speed of 130 km/h ≙ 800 Pa and safety factor of 3

The right is reserved to make technical modifications.

Qualifications

The ASE-300-DG-FT Module complies with the requirements of IEC 61215, NREL IQT, UL-1703 (USA) Fire Class A, Electrical Protection Class II and the EU guidelines, e.g. EMC according to DIN EN.

Figure 2.18: Modules specifications are found on module data sheets. Designers and installers should become familiar with the electrical and other characteristics in them (Source: schott.com)

2.4.9 The question of efficiency

The figures given for the efficiency of cells and modules are actually of little practical importance. Modules with lower efficiencies will simply cover a greater area of the roof – which is not a problem if there is sufficient suitable roof area. System costs are estimated not in terms of array surface area, but in terms of system peak power, expressed as € per kWp or US $ per KWp. It can also be expressed in terms of the energy yield of the system over a period of time, e.g. € per kWh produced over 20 or 25 years of operation. Modules themselves are compared in terms of € per Wp or US $ per Wp. Lower efficiency simply means that an array consisting of modules of lower overall efficiency will have a greater surface area and modules of different efficiencies can be realistically compared to each other. Photosynthesis, the most common use of solar energy on the planet, is *only* 1–2 % efficient. (See also 2.5 *Estimating the output of PV systems, array angles and orientation*, 3.2.7 *Inverter technical specifications and efficiencies* and 6.5.1 *Sizing stand-alone PV arrays*).

2.4.10 Quality requirements, quality marks and seals, standards and certification

The International Electrotechnical Commission (IEC) is the international standards and conformity assessment body for all fields of electrotechnology. The CEC (Commission of the European Community) Joint Research Centre in Ispra, Italy, has developed a testing procedure for PV modules. The resulting standard, CEC Approval Specification No. 503, has become the basis for the IEC standards for PV modules – IEC 61215 for crystalline modules and IEC 61646 for amorphous modules. The American Society for Testing and Materials (ASTM) is another source of standards. In the USA, Underwriters Laboratories or UL set standards for, and certify PV modules to UL 1703. Modules which comply with standards developed by these and similar organizations can be considered reliable and likely to have long working lives.

The most important current standards are:
- IEC 61215: Crystalline silicon terrestrial photovoltaic (PV) modules – Design qualification and type approval
- IEC 61646: Thin-film terrestrial photovoltaic (PV) modules – Design qualification and type approval
- IEC 61730: Photovoltaic (PV) module safety qualification – Part 1: Requirements for construction and Part 2: Requirements for testing
- EN 50380: Datasheets and nameplate information for photovoltaic modules.

IEC 61215, as well as a series of ASTM standards (incl. ASTM E 1038, 1171, 1596, 1802 and 1830) cover all the external conditions to which modules are exposed, such as mechanical stresses, climatic conditions and aging. Material and design guidelines regarding safety requirements are specified in Part 1 of IEC 61730. Part 2 describes the relevant test procedures. (See 7.2 *Standards for PV modules, system components and systems*).

It is difficult for the non-expert to judge the quality of a PV module. Quality marks and seals are useful. On the other hand, the lack of a quality mark or seal does not necessarily mean that a product is sub-standard. Smaller manufacturers, who actually produce good-quality modules, are sometimes discouraged from obtaining them by the expense entailed.

2.4.11 PV module warranties, guarantees and replacements

Warranties (called guarantees in some countries) are extremely important. There are essentially two types and it is important to differentiate between them. There are statutory warranties, and there are supplementary voluntary warranties offered by manufacturers and suppliers.

In most countries in North America and Europe, consumer protection legislation imposes a statutory obligation on manufacturers to provide warranties against product defects – these usually cover two years. The specifics as to when and the under what conditions these apply, and who has the legal obligation to honor them – the manufacturer or the supplier? – will depend on the relevant legislation. These statutory warranties may also place certain obligations on the purchaser, for example, that defects be reported within a specified time period.

In addition to fulfilling these statutory obligations, manufacturers can also offer supplementary warranties regarding particular aspects of a product. For PV modules this kind of warranty is particularly important. A typical warranty of this sort guarantees that after 20 or 25 years the output of a module will be 80% of its original rating. For example, that a module rated at 200Wp ±5% will produce 152Wp (200Wp -5% – 20%) after 20 years of use.

The small print will specify in detail which eventualities are covered and how claims can be made. The exact circumstances under which modules will be replaced or financial compensation offered need to be read. The small print needs to be gone through carefully and vague points and ambiguities clarified in writing. These supplementary warranties, not being statutory, are negotiable. They can be changed and improved. It is important that the conditions under which they apply are clear – for all parties concerned – to avoid difficulties in the event of problems. Ironically, one potential problem arises from the lengthy period covered by these warranties. It is quite possible that the module concerned will no longer be available, or even one with the same physical dimensions and electrical characteristics. It might have gone out of production because of improvements in the technology. A warranty is only as valuable as the capability of the guarantor to honor it. If the manufacturer is no longer in business and no longer in a position to honor it, even the best warranty is useless. So it is important that the manufacturer has adequate insurance to cover this eventuality. Suppliers should be asked about this. This type of insurance ensures the on-going validity of warranties and will protect both installer and customer. However, it must be said that the warranty rate for major PV manufacturers has been dropping steadily for years as they have learnt how to make better and better modules. The chances of ever encountering a defective crystalline silicon module from a major manufacturer are exceedingly small. One major manufacturer in the USA, who passed over $20 million worth of PV through their warehouse in 2005, had exactly one (1) defective module!

2.4.12 The cost of PV modules

At the time of writing, the installed cost of PV modules works out at approximately € 3,500 − € 4,500 (before tax) (€ 1 = US $ 1.2 approx., December 2005) per kWp installed, depending on module size, manufacturer and quantity purchased. It is expected that this price will decrease with increasing mass production. But prices can also go up, as they did in 2005 − despite increasing production − due to the increased demand for modules. This price increase was due mainly to the fact that, though module assembly capacity had increased, there were problems with the supply of silicon of the required purity. However, new silicon production facilities are due to come on line in the middle of 2006. The problem with the supply of high quality silicon is not a question of its being a limited resource. The raw material, silicon-oxide, is literally, as common as sand on a beach.

Figure 2.19: In 2005 the cost of modules was about € 3,500−€ 4,500 (€ 1 = US $ 1.2 approx.) per kWp installed, depending on module size, manufacturer and quantity purchased (Source: SunTechnics)

2.5 Estimating the output of PV systems, array angles and orientation

Solar radiation is measured in units of solar energy falling on a horizontal surface over a period of time – the standard unit is kilowatt-hours per square meter (kWh/m²). This is usually given for an average year or for an average day in a given month. These units are known as *peak sun hours* or *full sun hours*. To relate these units to the output of a PV module is quite straightforward. Under Standard Test Conditions (STC) the output of a PV module is

Wp (module peak power) × number of peak sun hours = kWh

For example, a PV module rated at 80 Wp which has 1,200 peak sun hours falling on it over a year at STC will produce: 80 Wp × 1,200 peak sun hours = 96,000 Wh or 96 kWh (see also 2.3.2 *Peak Power*). However, this is for ideal conditions which in reality only occur in the laboratory. In a real grid-tied PV system one could expect perhaps 75 % – 80 % of that, while in a stand-alone-system about 60 – 70 % could be expected, depending on circumstances. This difference between the actual energy yield or output of a PV system and the theoretically possible energy yield is known as the performance ratio.

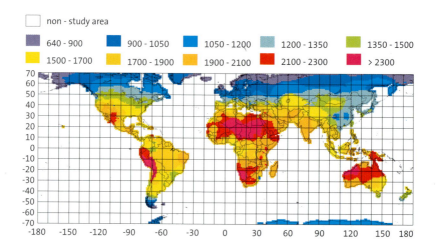

Figure 2.20: *This map shows the total solar radiation falling on a horizontal surface measured in units of total kWh/m² per year. However, maps like this are of limited use because solar modules are almost invariably not laid out horizontally but tilted towards the sun*

The information needed when estimating the output of a module or array is:
- solar radiation data for the longitude and latitude of the site
- the orientation of the module(s) (south, east, west?)
- the tilt angle or angle of inclination of the module(s)
- the peak power of the array (Wp)
- the performance ratio of the system.

Solar radiation consists of diffuse radiation and direct radiation. PV modules are sensitive to both. To predict exactly how much solar radiation is going to fall on a particular surface over a given period of time is not possible. Solar insolation levels vary from year to year. And because the sun is always moving across the sky, taking a different path each day, the mathematics of estimating how much solar radiation is going to fall on a PV array is not straightforward. The most accurate method of estimating solar insolation (and thus PV system output) is to use sizing software (see 2.6 *Sizing and design software),* but good results can also be obtained using manual calculations and very often that is sufficient, particularly for smaller systems. Also, different approaches are taken when working with grid-tie and stand-alone. In both cases there should be no array shading (see 3.10 *The problem of shading.)*

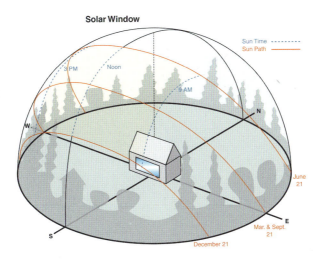

Figure 2.21: A three-dimensional illustration of the movement of the sun at a site at 40° North (Reprinted with permission of Home Power magazine, ©2005, www.homepower.com)

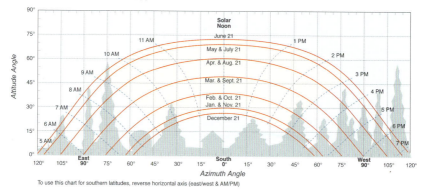

Figure 2.22: A two-dimensional sun path chart giving the same data
The maximum height (zenith) of the sun is different every day, and the point at the horizon
(azimuth angle) at which it rises and sets also changes daily (Reprinted with permission of
Home Power magazine, ©2005, www.homepower.com)

The importance of the angle of inclination angle and orientation

By tilting the angle of the module or array away from the horizontal, one can increase the quantity of solar radiation falling on it and increase the output of the array. In fact, ideally, the PV array should be kept at an angle perpendicular to the sun's incoming rays, but in reality most arrays are fixed at a particular angle. By choosing the correct angle and orientation, it is possible to optimize the output of a PV array for its particular application. However, when installing modules on roofs – which is more often the case with grid-tied systems than with stand-alone systems – angle and orientation are usually fixed. Just how crucial angle and orientation are depends very much on the diffuse solar radiation component, i.e. that component of total solar radiation coming from different parts of the sky. In more southern regions and in deserts there is more direct solar radiation, while in more northerly regions the diffuse component can be relatively high – for example, in Central Europe it is about 50 %.

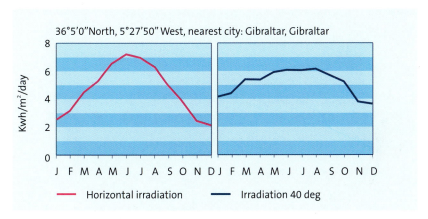

36°5'0"North, 5°27'50" West, nearest city: Gibraltar, Gibraltar

Kwh/m²/day

—— Horizontal irradiation   —— Irradiation 40 deg

Figure 2.23: The above graphs show the average daily global solar radiation for Gibraltar, Spain, 36° North, month by month. 1 On the horizontal plane. 2 On a plane tilted at 40°. Note that by changing the angle from the horizontal, solar insolation falling on a PV array can be changed – in this case decreased slightly in summer, and increased in winter. In general grid-tied systems work best at shallower angles – harvesting as much sun when it is sunniest, i.e. in summer; while stand-alone systems wich need to work through the year will have steeper angles – so that they can get as much sun as possible in the winter months when the sun is lowest in the sky (Source: PVGIS EU Joint Research Centre, http://re.jrc.cec.eu.int/pvgis/pv/)

Module support structures which track the sun can result in 30% to 40% more power output yearly and are a very good idea in some situations. Output is not only increased but is reasonably uniform – which can be an advantage in water pumping systems. However, they entail extra cost and moving parts can break down. In practice, most PV arrays are fixed.

Maps and bar charts

There are several types of solar radiation maps, and they can be broadly placed into three categories:
- maps showing total solar radiation in kWh/m² falling on a horizontal plane – of limited use as stated above
- maps showing total solar radiation in kWh/m² falling on a plane tilted at an angle of latitude – these are usually either in sets of four (each season) or twelve (for each month); they are useful for stand-alone systems but also have their limitations
- maps showing total solar radiation in kWh/m² falling on a plane tilted at a specific angle – these are the most useful – these can be generated manually but are more easily and accurately produced using computer software.

The information in maps can also be expressed for a particular site in the form of bar charts. These have the advantage of the solar radiation data being expressed on a month by month basis which is more useful for sizing stand-alone systems.

Average Daily Solar Radiation Per Month
Annual

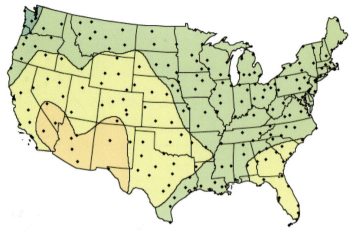

Flat Plate Tilted South at Latitude

Average Daily Solar Radiation Per Month
June

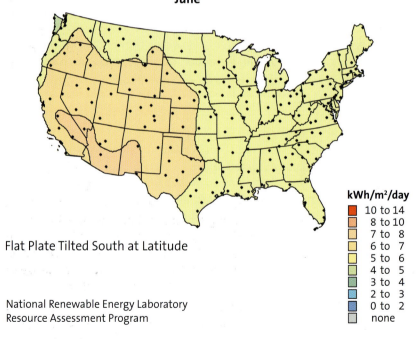

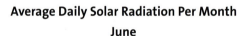

kWh/m²/day

■	10 to 14
■	8 to 10
□	7 to 8
□	6 to 7
□	5 to 6
■	4 to 5
■	3 to 4
■	2 to 3
■	0 to 2
□	none

Flat Plate Tilted South at Latitude

National Renewable Energy Laboratory
Resource Assessment Program

Figure 2.24: Maps of the USA showing total solar radiation in kWh/m² falling on a plane tilted at an angle of latitude (latitude tilt). 1 Showing average annual kWh/m²/day, 2 Showing average kWh/m²/day for June (Source: NREL, National Renewable Energy Laboratory, http://www.nrel.gov)

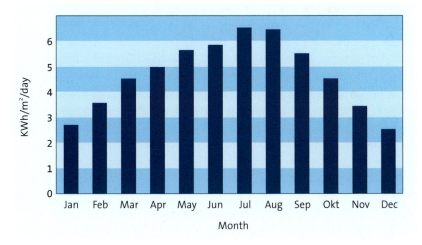

Figure 2.25: This bar chart gives an estimate of monthly average daily solar radiation at the angle of latitude 47° approx. (or latitude tilt) for a location in Montana, USA. (Based on solar insolation maps available on National Renewable Energy Laboratory (NREL) website – http://www.nrel.gov)

One of the major limitations of maps is that they do not take the orientation of the solar modules into account. This is important when the solar array is being installed on a roof, as roofs are rarely orientated exactly due south. Also, there can be very local variations in the quantity of solar radiation falling at any given site, for example, on opposite sides of a valley, or between a site on the coast and one a short distance inland. These local variations do not show up on any maps or even in software sizing system solar insolation data.

Estimating solar yield for grid-tied PV systems

When estimating the output of a solar array for a grid-tied PV system, the designer is usually faced with the problem of estimating what the annual output would be from a specific size of array. The size of the array is usually determined by budget constraints and physical constraints. And, if it is to be installed on a roof, the angle of inclination and orientation will also be predetermined. The easiest way of estimating output is undoubtedly by using sizing software (see 2.6 *Sizing and design software*). There are also web sites which will give output figures for arrays at a range of angles and orientations (see 7.6 *Sources of further information*).

A common and useful approach is first to ascertain what the expected annual output (kWh/kWp) of an array at the ideal angle for the latitude of the site would be. This information can be worked out mathematically, but the process is unwieldy and time-consuming, and it is more practicable and less subject to error to work with expected system output figures based on experience. Government and non-government web sites promoting solar energy and PV manufacturers will also have information in this regard. If the roof is not at the ideal angle or facing due south, which is more than likely, a *solar yield diagram* can be referred to. These enable the designer to estimate approximately what the deviation from the yield will be over the full range of roof inclinations and orientations. The solar yield diagram reproduced here is for a location in Central Europe where the optimum angle for grid-tied systems is 35°.

EXAMPLE: A roof in Berlin at an angle of 35° and orientated due south can be expected to produce on average 800 kWh per kWp of array over a year. So suppose, a 5 kWp PV array is being installed on a roof with 30° inclination. It is not facing due south – it is oriented 45° SW. What would the expected output be? On the ideal roof the yearly yield would be 800 kWh/kWp, so 5 kWp should give a yield of 5 × 800 = 4,000 kWh/year. The solar yield diagram now allows us to adjust that for the actual angle of the roof and its orientation. The yield on the roof can be expected to be 95 % of that, i.e. 3,800 kWh/year. One can also see from the diagram that tilt angles are not always critical and neither is some deviation from due south. In this example, any roof with an angle between 20° and 55°, and not deviating more than 30° from due will give 95 % of optimum output. This is because the diffuse component of solar radiation is relatively high in Central Europe. It is also obvious from the diagram that installing PV arrays on north facing roofs is not a good idea.

An alternative approach would to be use the figure for total solar radiation (kWh/m²/year) at the site and adjust that for angle and orientation using a solar yield diagram. What the actual output of the system will be, would depend on the performance ratio of the system, its overall efficiency.

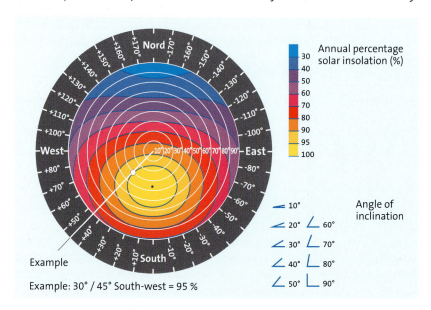

Figure 2.26: Solar yield diagrams are a very useful tool for estimating approximate outputs of grid-tied PV array on roofs. They are also available from some module manufacturers (Source: Ecofys)

Free-standing PV arrays for grid-tied systems and those on flat roofs (depending on circumstances) can be inclined and orientated at more or less ideal angles. Grid-tied PV arrays are usually tilted at shallower angles than those for stand-alone systems. In grid-tied systems the object is to generate as much solar energy as possible over the year, so they are pointed at the sun in the summer months when the sun is usually high in the sky.

Estimating solar yield for stand-alone PV systems

The approach taken with stand-alone systems is different for two main reasons. Firstly, the idea is not to generate as much electricity as possible annually, but rather on a daily basis to power the loads in the system. In practice this means that the required daily energy requirement is estimated for the least sunny month, which is usually in winter, and then the number of modules needed to supply that requirement is chosen. Usually, these systems are in remote rural locations and very often mounted on free-standing structures, so installing modules at the optimum orientation and angle is possible. In northern and southern latitudes, the angle of the array usually needs to be quite steep, which makes them unsuitable for mounting on a roof in any case.

One of the first things to do is to choose the optimum angle of the array. For locations north of the tropics, i.e. above 23° north (or south) a *panel angle graph* can be used. When the angle has been chosen, it is then necessary to find out what the daily average solar radiation is during the design month. Maps showing total solar radiation in kWh/m² falling on a plane tilted at an angle of latitude are a useful source of data. They will not be that accurate but very often accurate enough – particularly for small systems. Data can also be obtained at some of the web sites previously mentioned.

A very good source for basic solar radiation data is the NASA Surface Meteorology and Solar Energy Data Set web site. It can be found at http://eosweb.larc.nasa.gov/sse/. The site gives monthly average kWh/m²/day on a horizontal plane for any location on the globe. Longitude and latitude can be entered as figures or by clicking on a map. However, the figures need to be adjusted using correction factors to take tilt angles and orientation into account. These correction factors come in tables and are not always easily available. Each factor is for a specific tilt/orientation, a specific month and a specific latitude. For example: if a tilt correction factor for a specific month, specific tilt angle and specific latitude is 1.5 and the solar insolation data for a horizontal plane for that month is an average 3,500 kWh/m²/day, then the adjusted figure for the chosen angle will be 3,500 kWh/m²/day × 1.5 = 5,250 kWh/m²/day. (However, the situation is simplified somewhat by the fact that PV arrays on stand-alone systems are not generally mounted on roofs, so they are easily orientated due south, so correction factors for orientation usually do not need to be applied. It is an accurate way of obtaining solar data but it is rather cumbersome and not always necessary (see also 7.6 *Sources of further information*).

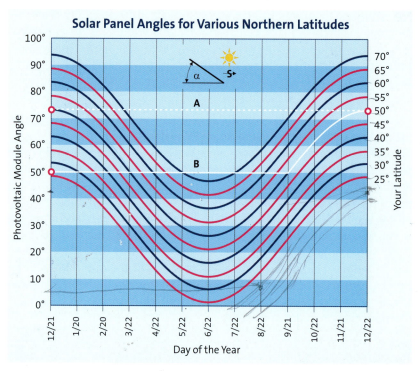

Figure 2.27: Panel Angle Graph: Using this graph it is very easy to choose the optimum angle for a PV array in a stand-alone system.
Example A: a bus shelter lighting system at 50° north. The design month is December, that is when there is the least sun and the most demand – the nights are long and dark. The optimum angle is between 70° and 80°.
Example B: a holiday cottage at 50° north to be used only between March and September. The design month can be either March or September. The winter months are not relevant. The optimum angle is 50°. Between 45° and 55° would be fine.
(Reproduced with permission of Home Power magazine, ©2005, www.homepower.com)

Orientating a module in the tropics is different. At the equator the sun is sometimes in the north, sometimes directly overhead and sometimes in the south. The general rule is to orientate the modules in the direction in which the sun is located during the least sunny months. This will balance the average daily yield over the year. The modules should not be placed flat. The angle should be about 20° – this ensures that they are cleaned by the rain.

When estimating the output of PV modules in stand-alone systems with batteries, amp-hours (Ah) are also used, rather than (Wh). This is because batteries are sized in Ah not Wh. See also 6.4.4 *Assessment of solar resource for stand-alone PV system sites* and 6.5.1 *Sizing stand-alone PV arrays*.

2.6 Sizing and design software

Numerous computer programs are available which enable one to size arrays, to estimate array outputs, to size components, cost systems and provide the cost of units of energy produced and even provide layout and configuration options for modules. For designers doing large systems these are essential. However, simulation software is useful only to the extent to which they can model real situations. One needs to learn how to use them and this takes some time. When first using a simulation software package, support from the software provider is necessary. A thick manual and online help is not enough, one needs to be able to talk to someone on the telephone in normal business hours. Some programs and the designing and sizing of non-standard systems require a high level of knowledge on how PV systems work. (See 7.3 *Sizing and design software* for some commercially available programs.)

Sizing programs are also a good marketing tool. Most will produce attractive graphics showing system output and system design and going through the design process with the customer can also be useful. For projects in receipt of public funding a computerized simulation may be required by the funding body. For installers carrying out regular installations, computerized sizing and design programs are well worth the money and the time involved in learning how to use them.

Good sizing and design software offers the following advantages:
- time spent on sizing and planning can be reduced
- several options can be tried out
- systems can be designed with greater precision
- tailoring system size and design to the customer's needs is easier
- income projections for grid-tied systems can be calculated reasonably accurately.

The way the software is used depends on whether one is designing a grid-tied system or a stand-alone system. When designing a grid-tied system, the orientation, roof angle and location is entered, followed by the type and number of modules and the inverter details. The program then calculates the expected annual output from the array. Some programs will even run plausibility checks. When designing a stand-alone system, the energy requirements are fed in first and the program selects the number of modules, the size of the batteries, inverter and charge controller. With stand-alone systems an estimate of down-time/coverage is useful. It is impossible to guarantee that systems will work 100 % of the time, even utility power cannot be guaranteed 100 % of the time.

Care needs to be taken if the result of a computer simulation is included in a contract. For example, the output of a large grid-tied system might turn out to be lower than that predicted by the sizing software, giving a lower expected income from the system.

Besides commercially available simulation software, some manufacturers offer sizing software which is free or very cheap. However, manufacturer-specific sizing software is usually designed only for systems using that manufacturers' products. This may not give optimum results.

Some manufacturers and suppliers will be happy to run though system design on their own sizing and design software. In any case it is always a good idea, especially for installers working on their own, to check designs out with whoever is supplying the equipment.

The data bank of the simulation software will contain the technical specifications of a range of modules, inverters and other components, but there are new versions coming on the market all the time, so the data bank has to be kept up to date. Some software providers provide regular updates. With some, details of new products can be entered. Sizing software will have standard radiation data, and it is usually not a problem using this. However, there may be local variations which may be significant – some software will allow this more accurate data from meteorological services to be used.

Experience has shown that yield predictions can be quite accurate when the actual installation is installed and operated the way it was designed and planned in the simulation program. However this is not a guarantee, so experience needs to be taken into account and results checked – perhaps with a system supplier with relevant experience. Special care needs to be taken when including the results of simulations in contracts. Agreements with customers should state clearly that simulations are simulations and that any predictions they contain regarding system performance are predictions, not guarantees. It should also be kept in mind that solar energy varies ± 10 % yearly, so hair-splitting estimates just aren't necessary.

2.7 Purchasing PV modules and other system components

PV systems are usually designed on the basis that the modules will last at least 20 years. Whether the system will last that long or not depends on the quality of the modules and the other system components. Batteries in stand-alone systems may need to be replaced every few years. Very often the more expensive installation is the better one. The attitude that solar modules simply lie in the sun and produce electricity can be taken and can work, but the professional installer should pay close attention to module quality and warranties. Here are some general tips on things to look out for when purchasing PV modules and other PV system components:

- the performance warranties and what happens if there are problems – some manufacturers and distributors will simply offer replacement modules with the installer having to bear the cost of dismounting the faulty module, installing the new module as well as freight etc. – others will undertake to do all of that with their own staff
- module replacement – will it be the exact model or merely one of the same wattage and dimensions?
- the fine print – some products need regular maintenance in order for the warranties to remain valid, e.g. batteries, some grid-tie inverters
- the ability of the manufacturer or the system supplier to honor warranties is important
- when buying complete systems – is the supplier purchasing the individual components from reliable sources or simply buying the cheapest on the market at the time?
- the speed with which suppliers can replace faulty equipment – for example a faulty inverter on a large grid-tied system can mean loss of income for the system owner
- how often equipment may need to be serviced – codes may specify periodic inspection and testing, batteries in stand-alone systems usually need regular maintenance
- manuals can be downloaded from websites and read before purchase of an item of equipment – inadequate manuals are not a good sign
- compliance with national electrical and building codes.

3 Grid-tied PV Systems

3.1 Principal components of a grid-tied PV system

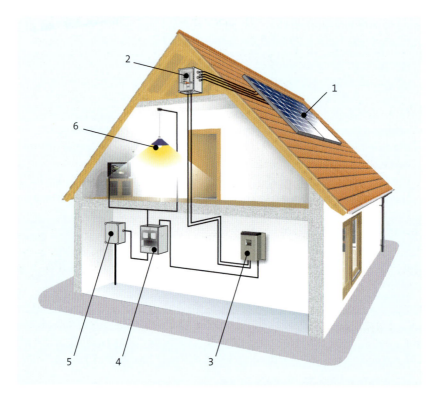

Figure 3.1: 1 PV array, 2 PV array combiner/junction box, 3 Grid-tied inverter, 4 Import-export meter, 5 Connection to the grid, 6 Loads. Other configurations are possible

In a typical grid-tied solar electric system, the DC electricity produced by the PV array is usually fed by cables into a *PV array combiner box* (sometimes called the *PV array junction box)* where they are connected together. A cable from this junction box feeds the DC electricity to the grid-tied inverter. The inverter converts this DC electricity into AC which is either consumed by the building loads and appliances or fed onto the grid. The inverter is either connected to the main AC circuit breaker panel/fuse box or directly to the incoming cables from the grid (see 4.6 *Connection to the grid and meter location* for possible configurations). If the PV array is not supplying enough electricity to power the loads in the building (for example at night), electricity is supplied by the grid. An export meter meters the amount of electricity that the system puts onto the grid, and an import meter meters the amount of electricity imported from the grid. Usually it is one single meter called an import-export meter or, in the USA, *net-metering*. The whole process is automatic and seamless. Power flows back and forth like water seeking its own level.

3.2 Grid-tied inverters

Grid-tied inverters convert the DC electricity produced by the PV array into single-phase or three-phase electricity at a voltage and frequency suitable to be fed onto the grid. They are available in a range of sizes and are rated in Wp, the peak wattage of the PV array they are connected to. They are also called *grid-connect inverters* and *utility-intertie inverters* or *synchronous inverters*.

Inverters for use in grid-tied systems are not the same as those used in stand-alone systems. The terminology is sometimes confusing as some stand-alone inverters (inverter-chargers) can also be connected to the grid – but only to import power from it, not export it (see 6.3.5 *Inverters in stand-alone systems*).

Different types of inverters are available for different PV array configurations: central inverters which will serve a single installation, inverters for single strings of modules, inverters for multiple strings of modules and inverters for single modules.

In smaller single-household systems, the PV array is usually connected to a single inverter. However, if the array is large or if there is unavoidable partial shading of the array or if parts of the array are orientated in different directions or have different inclinations, several inverters may be required.

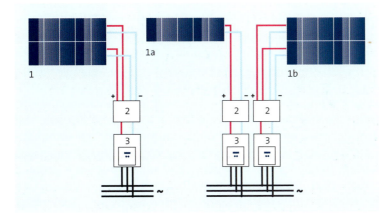

Figure 3.2: An installation with a central inverter (left) and one with several inverters (right). 1 PV array, 1a & b PV array divided into two parts, 2 PV combiner box, 3 Inverter

The principal functions of a grid-tied inverter are:
- to convert the DC electricity produced by the PV array into AC and feed it onto the grid at the required voltage/frequency/phase – 120/230 VAC/60 Hz in North America, 240/400 VAC/50 Hz in Europe.
- to maximize the output of the PV array – under the varying conditions of solar insolation in which it will operate – by tracking the point of the I-V curve of the array at which electrical power output is maximized: this is known as maximum power point tracking (MPPT)
- to ensure non-hazardous operation of the system by complying with all relevant electrical codes and automatically disconnecting itself from the grid if the grid is de-energized, or strays out of strict voltage/frequency limits (see *islanding* below).

Inverters are available in a range of sizes and there are different types. Some will have transformers and some not. National codes need to be referred to regarding which types of inverters are permitted in each country.

Grid-tied inverter technical requirements
The main technical requirements of a grid-tied inverter are:
- generation of a pure sine wave synchronous with the sine wave of the grid
- accurate tracking of the MPP of the array I-V curve
- high efficiency operation at full and part loads
- automatic operation
- reliable performance in conditions of both high or low ambient temperature
- visual display of array output, fault indicators etc.
- compliance with national codes and regulations.

Inverter grid monitoring and *islanding* prevention
The inverter also needs to disconnect itself automatically from the utility grid if the grid is turned off by the utility in order to carry out works. The situation in which the grid has been turned off by the utility and the inverter keeps putting a voltage on the grid is known as *islanding*. This must be prevented from happening – as it would present a serious hazard for utility electricians working on the grid at that time. Achieving this is complicated by the need to ensure that, in an area where there are several PV systems and the grid goes down, one PV system does not interpret voltage put onto the grid by another PV system as evidence of the grid still being generally energized. The electronics ensuring that this never happens are sophisticated. Grid-tied inverters are very sensitive to variations in grid voltage, frequency and impedance and will only operate within a certain range – outside this range they will automatically shut down (see 5.4.2 *Problems originating on the utility grid* for more details). Sometimes utilities may insist on carrying out their own tests to ensure compliance with their requirements, but testing and certification of inverters at the factory may be acceptable. UL1741 listing is an automatic approval in the

USA. Codes and regulation covering this aspect of grid-tied inverters are very strict and need to be referred to and utilities should always be consulted. The sensitivity of grid-tied inverters makes their use in situations where there is an erratic grid – such as in many developing countries – impracticable.

3.2.1 Grid-tied inverters and PV module configurations

Modules can be connected in series, in parallel or in series-parallel. The choice of inverter(s) to match the wattage of the array determines string sizing for the PV modules.

Inverters can be categorized in several general types:
- *Central inverters* to serve a whole installation, starting at 5 kW
- *String inverters – single-string inverters* which are connected to one string of modules and *multi-string inverters* which are connected to two or more strings of modules (0.7 kW – 5 kW)
- *Module inverters* for one or two modules – sometimes module integrated (0.1 kW – 0.7 kW), (not available in the USA at the time of writing).

When modules are connected in series the output current of the string remains that of the individual module (in fact, the lowest) while the output voltage is the sum of the voltages of all the modules in the string. If modules are connected in parallel, the overall output voltage remains the same and the current is the sum of all the module currents. Connecting the solar modules in series (where there is no array shading) has the following advantages:
- installation is faster and less complicated
- the increased voltages produced allow the use of cables of a smaller cross-section size – important where cable runs need to be long (see 3.3.1 *Cables*).

In grid-tied systems modules are usually connected in series strings. The maximum voltage of a string of modules (the sum of the maximum voltages of all the modules in the string) must be lower than the maximum input voltage rating of the inverter. However, modules may need to be connected in parallel if part of the array is shaded or to prevent *mismatching* in strings if there are significant variations in module electrical characteristics or to keep within required voltage levels.

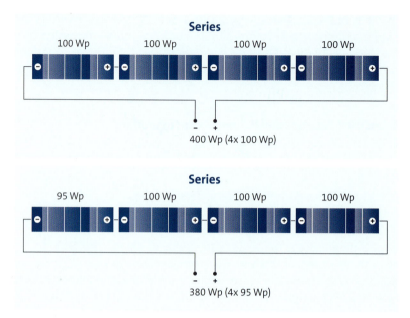

Figure 3.3/3.4: PV modules connected in series: the output current of the string remains that of the individual module (in fact, the lowest) while the output voltage is the sum of the voltages of all the modules in the string. Wiring a module with a lower current output than the others in a series string is an example of mismatching

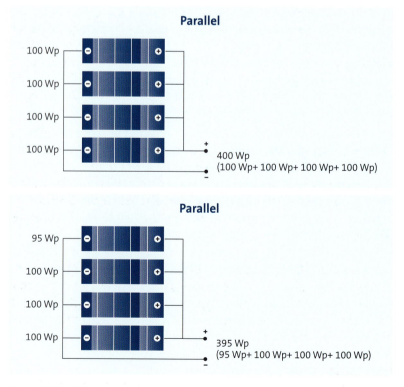

Figure 3.5/3.6: PV modules connected in parallel: the overall output voltage remains the same and the total current is the sum of the module currents

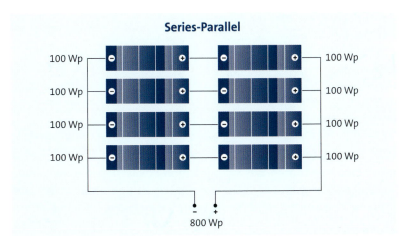

Series-Parallel

100 Wp
100 Wp
100 Wp
100 Wp

100 Wp
100 Wp
100 Wp
100 Wp

800 Wp

Figure 3.7: PV modules connected in series-parallel

EXAMPLE 1: The nominal performance $P_{NOMINAL}$ of a module is 200 Wp and its open circuit voltage (V_{oc}) is 72 V (at -10 °C). The inverter has a maximum DC input voltage of 600 V. How could a 1.6 kWp array be assembled? 8 × 200Wp modules would give 1,600 Wp or 1.6 kWp. If they were connected in series (i.e. in one string), this would give a string output voltage of 8 × 72 V, a total of 576 VDC. This is below the maximum DC input voltage (600 VDC) of the inverter, which is acceptable.

EXAMPLE 2: The same modules are to be used in a 3.2 kWp array. This would require 16 modules – 16 × 200 Wp = 3.2 kWp. Connected in series they will give 16 × 72 V = 1,150 VDC, which is above the maximum input voltage of the inverter. A solution here might be to connect the modules in series-parallel – two strings of 8 modules connected in series. This would give a maximum voltage of 575 VDC, which is acceptable.

3.2.2 Inverter design concepts – central inverters, single- and multi-string inverters and module inverters

There is no single hard and fast rule for what inverter design concept or configuration to select. The specifics of the installation and the compatibility of system components have to be accessed before arriving at an optimal solution.

Central inverters – in conjunction with higher array operating voltages

In this type of installation all the modules in the array are connected to one single inverter. Usually, the array will consist of several strings which are then connected up together in the PV combiner box before being connected to the inverter. The principal advantages are:
- high array outputs possible – up to the megawatt realm
- robust construction
- the termination of all the DC cables in the PV combiner box is relatively straightforward.

The principal features of central inverters are:
• centralized installation
• combined series and parallel connection of modules, though mainly series
• suitable only where the PV array is subject to a uniform regime of solar insolation, not where there is shading or parts of the array are orientated in different directions or have different inclinations
• only suitable for modules which have the same electrical characteristics.

If strings of modules of different sizes (Wp), types, and power tolerances are connected together in the main junction box, a single array voltage and current will be produced. If the inverter has a single input, it will read the overall voltage and current produced by the array and track the MPP of the array accordingly. It will be unable to differentiate between the different strings which have different MPPs. The result is that the efficiency of the installation will be much lower than if a multi-string inverter, which could differentiate between the strings, was used. The same problem could arise with a shaded string.

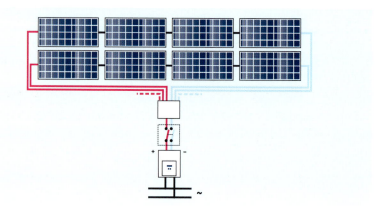

Figure 3.8: The PV array consists of several strings of series-connected modules. The whole of the installation is served by a single central inverter

Central inverter – in conjunction with lower array operating voltages
In conjunction with lower array operating voltages the use of a central inverter has the following advantages:
• performs better where there is inevitable array shading
• more suitable for modules with larger power tolerances
• less hazardous voltages (< 120DC) possible.

The principal features of the configuration are:
- the array consists of several strings with a lower number of modules in each string
- the higher currents produced can require larger diameter cables
- inverter DC input voltages will be lower.

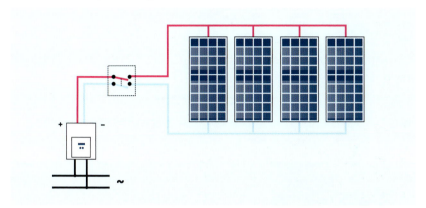

Figure 3.9: Solar modules connected in parallel to a single inverter. This configuration is advantageous where partial shading may occur or if the modules have high power tolerances

Single-string inverters

This type of inverter is fed by a single string of series-connected modules. Size can range from 0.7 kW to 2.5 kW. Larger installations will require several. The advantages:
- the MPP of individual strings are tracked, which leads to higher overall array output
- optimal choice where there is part shading of an array and/or the strings have different orientations/inclinations
- no need for a PV combiner box
- the inverters do not have to be installed in a central location – they can be near the strings – DC cable runs are thus reduced.

The principal features are:
- each inverter gets the most out of the modules it is connected to
- different conditions of solar insolation (caused by different orientations, inclinations and shading) in parts of the array are acceptable
- the strings are, in effect, connected together on the AC side, each string being essentially a unique grid-tied system
- each string should be composed of modules with similar power ratings, power tolerance and be subject to the same solar insolation
- used in bigger installations.

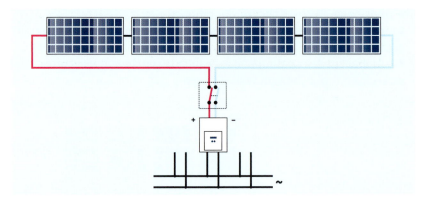

Figure 3.10: Single-string inverters take a single string of series-connected modules. Each string has its own inverter

Figure 3.11: This situation is not optimal. The two groups of modules on the roof have different inclinations, and are apparently different kinds of modules. A separate inverter should be used for each section of the array, or a multi-string inverter (Photo: Thomas Seltman)

Multi-string inverters

Inverters which are a combination of the single-string and central inverter concepts have now been on the market for a while. Their size is about 3 to 5 kW. On the array side, they consist of several single-string inverters but, on the grid side, they are like central inverters. They were developed specially for use in situations in which an array does not have a uniform orientation or inclination or consists of different types of modules or of strings of different numbers of modules or where there is shading. The advantages are:

- a single inverter can be used with strings which have different requirements, which is cheaper than installing several single-string inverters
- performance of each individual string is optimized by separate MPP tracking of each individual string, leading to a higher overall array output.

The principal features are:
- several integrated MPP trackers
- each inverter can be connected to two or three strings, each string with different orientations, inclinations and power/power tolerance ratings.

Module inverters

There are also small inverters for single modules or module pairs. They are mainly used in smaller installations, and not in the USA, but there can be significant advantages in using them in larger installations as well. The advantages are:
- no need for any DC wiring or cables
- shading of a module or a fault in a single inverter will not affect the rest of the array
- ideal when using modules of different power tolerances
- less hazardous voltages (< 120 DC) possible.

The principal features are:
- each module or pair of modules has its own inverter, which is optimally compatible with its performance and specifications
- the inverter can be factory-fitted into the module junction box or attached to it – so-called *AC modules*.

 Inverters have an expected shorter life than modules, and a dislike of hot environments. If a module-integrated inverter develops a fault, the module might also have to be replaced.

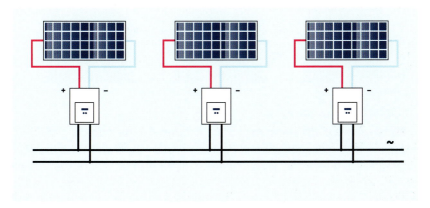

Figure 3.12: Module inverters connect single modules or pairs of modules directly with the grid. (Inverters drawn much larger than reality.)

3.2.3 Multiple interconnected inverters

Grid-tied inverters are at their most efficient when they are operating at or near full load. At reduced loads (i.e. under the conditions of reduced solar insolation), they will be less efficient. In an array consisting of several inverters, all of them would be operating at this reduced efficiency much, if not most of the time. However, if the strings in the array could be switched so that the array would in effect be fed through one inverter, then that would be working at a higher efficiency and array output would be higher.

EXAMPLE: In an installation consisting of several inverters, as the sun comes up, electricity is fed into the grid, not by all simultaneously, but by one inverter. When the solar insolation has reached a level at which one inverter is working at its full capacity, then another inverter is brought on line until that too is working at full efficiency, and so on until all inverters are on line. Similarly, as solar insolation drops in the evening, the inverters are taken off line one at a time, until there is only one working.

In order to do this, inverters need to be connected together so that they can communicate with each other. The working life of an inverter also depends on how often it is on. In this type of set-up, because the inverters are not on all the time, their real working lives are extended. The extra cost involved in inverter-switching is especially worthwhile in larger installations in which a small percentage increase of output can be significant.

3.2.4 Inverter- PV array compatibility, inverter size and location

Both inverter and PV array need to be compatible with each other:
• the size of the inverter (W) should never be less than 90 % of the peak wattage (Wp) of the array. Inverters are usually slightly under-sized. So, for a 10 kWp array an inverter of 9–10 kW can be used
• the inverter's MPP range (i.e. the voltage range between which it will track the MPP of the array, e.g. 350 to 600 VDC) must match the operating voltages of the array
• the inverter must be capable of withstanding the maximum array voltage and current.

If a complete system is being bought, the compatibility of the inverter with the array has usually been sorted out by the supplier, but it needs to be checked.

Inverters should never be undersized more than 10 % in a system in which the array is optimally oriented and at an optimal angle of inclination. The energy yield will be reduced and the life of the inverter will be shortened. Only in cases where the array will not reach optimum performance because of non-optimum orientation (an east- or west-oriented array, for example) or inclination (a vertical facade) can the inverter be undersized more than 10 %. Experience is needed in order to design systems like this.

Under no circumstances should the maximum voltage and current of the PV array / string exceed the input voltage and current ratings of the inverter. This can damage the inverter. Overloaded electrical components will also age more quickly and the working life of the inverter will be reduced.

When an inverter is overloaded or overheated it *derates* itself. This means that the inverter is no longer able to process a portion of the array power in summer, when it is working at its peak. Installing an inverter under the roof in a space which gets hot in summer can be problematic. At temperatures above 70 °C (approx.) most inverters begin to regulate the array output downwards in order to protect themselves. If it is not possible to avoid a hot location, an inverter with a ventilating fan should be chosen. Convection-cooled inverters (i.e. without fan cooling) will get too hot. Somewhere that remains cool throughout the year is ideal. Outside north walls are every inverter manufacturer's favorite. (See also 4.4.5 *Grid-tied inverter installation and wiring*)

3.2.5 Inverter technical specifications and efficiencies

The following specifications are needed in order to match the inverter to the array.

On the PV array/DC/input side:
- DC nominal power and DC peak power input
- DC nominal current and DC peak current input
- DC nominal voltage and DC peak voltage input
- the MPP voltage range
- minimum power required before inverter starts feeding into the grid
- stand-by power consumption.

And on the or the grid/AC/output side:
- AC nominal power output and AC peak power output
- AC nominal current output and AC peak current output
- inverter efficiency over a range of loads – at 5 %, 10 %, 20 %, 30 %, 50 %, 100 % and 110 %.

Inverters operate at different efficiencies, depending on the load. This is expressed as the inverter's efficiency curve. Different inverters will have different efficiency curves. Figures given for the efficiency of inverters under full load are not enough for planning purposes or comparing different inverters with each other, because for much of the time inverters will not be operating at full load – they will only do this mainly in the middle of very sunny days. The European Efficiency standard (valid for the type of irradiance levels found in Central Europe) is a method which enables different inverters with different efficacy curves to be compared by taking into consideration the amount of time the inverter can be expected to be operating at particular percentage loads/levels of solar insolation:

$$\eta_{EURO} = 0.03\ \eta_{5\%} + 0.06\ \eta_{10\%} + 0.13\ \eta_{20\%} + 0.1\ \eta_{30\%} + 0.48\ \eta_{50\%} + 0.2\ \eta_{100\%}$$

where the numeral refers to the proportion of time the inverter is operating at the percentage efficiency indicated, e.g. $0.2\ \eta_{100\%}$ means operating at 100 % efficiency at 0.2 percentage of total operating time.

Overall system efficiency η can be calculated as follows (the figures are just an example):

$$0.12\ \eta_{MODULE} \times 0.9\ \eta_{INVERTER} \times 0.99\ \eta_{CABLES} = 0.10\ \eta_{SYSTEM}$$

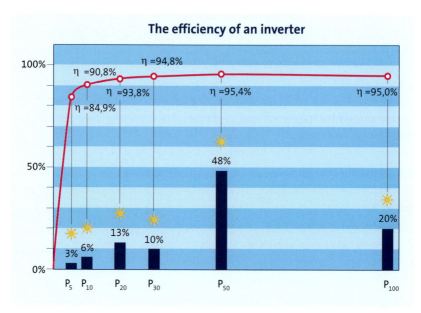

Figure 3.13: The inverter in this example has a European Efficiency of 94.5 %. The maximum efficiency is 95.4 %. It operates at this level of efficiency when the inverter is operating at 50 % of its nominal rating and it spends 48 % of its time at that highest level

Some inverters have transformers and some do not. Those without transformers will achieve higher efficiencies but they are not always compliant with national codes, and where they are allowed, there may be specific safety requirements.

3.2.6 Inverters – other issues

Inverter working lives

Grid-tie inverters are essentially reliable these days but the expected working life needs to be considered when assessing the overall economics of an installation. If it is shorter than that of the modules, it may need to be replaced. To date, working lives of between 10 to 15 years have been typical and now many manufacturers are designing their inverters to last 20 years and longer. If an inverter does stop working, the easiest and simplest solution is a complete replacement with a factory-reconditioned unit. The defective inverter can be taken back by the manufacturer and the faulty parts repaired or replaced without time pressure. The reconditioned inverter can then be used as a replacement in another installation.

Inverter prices
Current prices range between € 700– € 900 per kW (before tax) (€ 1 = US $ 1.2 approx., December 2005) . Similarly to modules, it depends on manufacturer and quantity purchased and on size. Like modules, the larger the inverter, the lower the price per kW.

Level of service offered by inverter supplier
The level of service the inverter supplier or manufacturer is able to provide is important:
- will a defective inverter be removed by a service team and replaced at the same time, or does it have to be delivered to them?
- are replacements readily available?
- when the warranty runs out, are fixed repair and servicing prices offered?

Hidden servicing costs
There can be hidden costs. With some inverters in some countries a functional test may need to be made every few years. The manufacturer needs to be asked about this and customers need to be informed about any extra servicing costs.

Weight
The logistics of replacing inverter/s in the future should also be considered. Inverters with transformers weigh about 10 kilos per kW. A crane might be necessary to bring a large inverter into position. Using several smaller inverters might be a better option. The location might also be no longer as accessible as it was during the original installation.

Inverter warranties, guarantees and replacements
Warranties are just as important for inverters as they are for modules. In addition to statutory obligatory warranties, manufacturers also offer further voluntary warranties. Some manufacturers offer 5 years, and up to 10 years on payment of an additional fee. Many of the points made in 2.4.11 *PV module warranties, guarantees and replacements* are relevant here. In the USA, all inverters carry at least a 5-year warranty as standard equipment, with optional cost extended.

Figure 3.14: Inverters should be installed in a cool location, these SMA inverters are installed on a north-facing outside wall (Source: SMA)

3.3 Other components in grid-tied PV systems

3.3.1 Cables

Cables used for the DC wiring of grid-tied PV arrays must have additional properties to those used in normal AC installations. They generally need to be:

- double-insulated (though not in the USA at the time of writing)
- UV and water-resistant
- rated for higher temperature ranges (approx. -40 °C to 120 °C)
- rated for higher voltages(≥ 2 kV)
- easy to work with, light and flexible (multi-stranded)
- flame resistant, low toxicity in case of fire, halogen-free
- sized for low voltage drops (usually 2 %, but see relevant codes)

Higher system voltages mean lower currents. This means that cable voltage drop and thus losses in the cables will be less. Power loss in a given cable is proportional to the square of the current flowing in it *(P = I2 R)*. Higher currents require cables with a larger cross-sectional area and add to costs. (See 7.7 *Power losses and voltage drop in PV systems*).

Suppliers can usually supply suitable cables. Pre-prepared cables with multi-contact water-resistant double-insulated and polarized (reverse-polarity-protected) DC connectors are fairly standard these days. They make the installation a lot easier. They can also be made up by system or cable suppliers to specified lengths.

3.3.2 The PV combiner box
If an array is composed of several strings, a PV combiner box is usually necessary to connect up all the cables which are to be fed into the inverter. Its principal functions are:
- connecting together – in parallel – the cables coming from the module strings
- facilitating testing of the module strings
- housing string fuses
- housing surge/over-voltage protection.

The function of string fuses is to prevent, in the case of a damaged DC string cable, an unacceptably high current occurring in other string cables. Surge/over-voltage protection is to protect against a lightning strike near the installation inducing voltages in the system which can damage the modules and inverter.

The PV combiner box needs to be double-insulated, usually have protection class IP 54 and enable the physically separate termination (non-conducting plates or junction boxes) of the clearly identified positive and negative conductors. If there are several inverters, a PV combiner box is required for each one. A PV combiner box may also house the main DC disconnect/isolator. This enables the inverter to be de-energized from the DC side. But if it does incorporate the main DC disconnect/isolator, it must be easily accessible.

Sometimes, in smaller installations, the DC cables are connected up in parallel using Y-plugs. However this practice has hidden costs, because the PV combiner box, as already stated, is used for more than just connecting the cables in parallel – it also houses string fuses and surge/over-voltage protection.

 National electrical codes need to be referred to when deciding whether to use a PV combiner box or not, and regarding the requirements for string fuses and surge / over-voltage protection.

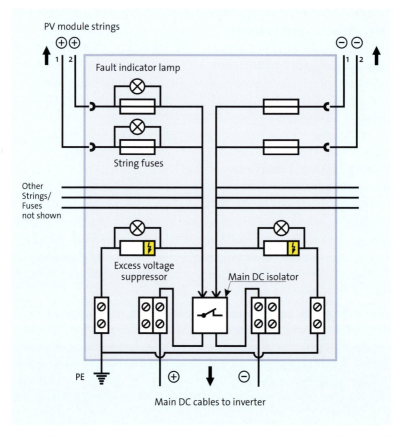

Figure 3.15: The main PV combiner box can house the main DC disconnect/isolator, string fuses and lightning/surge/over-voltage protection

3.3.3 The main DC disconnect / isolator

When it is exposed to light a PV array will always produce a voltage, so a DC disconnect/isolator is needed in order that the array can be disconnected from the inverter during installation, repair or maintenance. It must be capable of disconnecting the array from the inverter under full current. Ideally it is integrated into the PV combiner box, if this is easily accessible. In certain circumstances the PV combiner box may not be necessary, but the DC disconnect/isolator always is.

3.3.4 Meters and net-metering

Metering electricity sold onto the grid and purchased from it through a single meter, is known as net-metering. Meters will usually be installed by the utility and remain their property. Several configurations are possible (see 4.6 *Connection to the grid and meter location*). Metering requirements and options vary from country to country and utility to utility. Billing methods will also have an influence on the configuration and number of meters. Both the grid-operating company and the utility to which the electricity generated by the PV array is being sold need to be consulted (if they are different companies). In systems where most of the electricity generated is to be used in the building itself, the situation known as off-set, there may be no net-metering requirement. However, in this case it is recommended that a meter be installed to at least record the output of the PV array. The record of array output made by this meter can be indispensable in analyzing any problems which may arise.

3.3.5 Remote monitors

High-quality inverters incorporate remote data communication and data-logging features which allow the system owner to monitor performance via a laptop or PC, or website. Real-time performance, daily output, yearly output, system faults and malfunctions of inverters can be monitored. Wireless transmission of this data is particularly useful (see also 4.4.6 *Installing remote monitors*). A modem can enable it to be viewed at other locations. Many customers who have paid a lot of money for an installation are interested in monitoring system performance and showing visitors what the array has produced. It can look particularly good in the customer's living room. It is also an additional sales opportunity.

3.4 PV module mounting structures and systems

A vast number of module mounting structures and systems are available. The most common are structures which are actually fixed onto the roof. Other systems incorporate the modules into the roof itself or building facade. Free-standing structures for use on flat roofs or on the ground are also available, as are customized systems for individual installations. On some the inclination or tilt angle of the module can be adjusted – this can be useful on roofs with shallow inclinations. An important difference between systems is the actual layout of the modules. Portrait arrangement is preferable if the array is to consist of several rows of modules, strips of accumulated dirt above the lower edges of module frames are less likely and snow slides off more easily. Some of the main points to be aware of when choosing a mounting structure are:

- not all mounting systems are suitable for all modules, and some are specific to a particular make of module
- sourcing the modules and the mounting structure together makes sense – that way one can be fairly sure that it is all going to fit together
- there needs to be space under the modules to allow sufficient airflow
- relevant building codes and regulations need to be consulted
- the suitability of the roof for the mounting structure may need to be assessed and certified by a structural engineer
- a building planning permit may be required
- aesthetics are important – it needs to look good.

(See also 4.3.4 *Installing the modules – general guidelines*).

Mounting structure materials

Aluminum and stainless steel and are the most common and galvanized steel is acceptable. However, the integrity of the galvanization should not be compromised by on-site cutting or drilling. All screws, nuts and bolts should be stainless steel. Otherwise they can begin to corrode shortly after the structure has been installed. Local environmental conditions, such as pollution or proximity to salt water, need to be taken into consideration.

Figure 3.16: The choice of materials is important. Screws may need to be taken out after 20 years of exposure to the elements (Source: Tauber Solar GmbH)

Roof mounting structures

- The supplier of the mounting system should be provided with all the relevant details of the roof on which it is to be fixed.
- Space may need to be left between the modules for mounting clamps/fixings, depending on the system.
- Trunking or channeling for the DC cables should be part of the structure.
- The services of a roofer may be required.

Roof integrated mounting structures

- Systems which incorporate the modules into the roof are more time-consuming and require a greater level of skill to install. Involving the system supplier in the actual installation is advisable on the first occasion.
- Allowing for sufficient air flow at the back of the modules is particularly important. Heat build-up can reduce performance by up to 20 %.

- On flat roofs (roofs with an inclination of less than 27°) a structure which fits onto the roof is usually preferable from the point of view of maximizing output.
- Roof integrated systems are usually only available for portrait module layout.
- If the roof is unusual or special or otherwise of high value – an historic building for example – it is worth investigating options which blend with it.
- Roof incorporated systems also need to fit aesthetically around existing roof structures, such as skylights, dormer windows and possibly solar water heaters.
- The involvement of an experienced roofer is especially recommended.

Free-standing mounting structures, such as on flat roofs
- Not all roofs are suitable for free-standing mounting structures. The effect of the wind on the structure is one consideration – the modules act like sails – and the structures themselves are usually weighed down with aggregate or cement – an additional load on the roof.
- They need to be high enough off the roof (or ground) to avoid being buried in snow in winter or splashed with dirty water.
- Sufficient space needs to be left between the rows of modules so that the rows do not shade each other – especially in winter when the sun is low in the sky.
- Protection against theft is an issue – unauthorized removal should, at least, be very difficult.
- On high roofs exposed to high winds the integrity of a structure might need to be enhanced by fixing the rows of modules to several points on the roof – this should stop any movement or toppling over of the rows of modules.
- The services of a roofer should be engaged – holes may need to be made in the roof and sealed.

Facade mounting structures
- Mounting modules on the facades of buildings is a specialist area, not for the beginner.
- Building regulations for facades need to be referred to.
- Not all modules are suitable for mounting on facades.
- Co-operation with a builder who specializes in facades is strongly recommended.

3.5 Where to install: roof, facade or ground?

The advantages of roof mounting structures

- Roof mounting structures are the most straightforward and the speediest to install and the most economical method. They can more often be installed without outside help.
- They can usually be put onto existing roofs and older roofs
- Good airflow is provided at the back of the modules.
- There is minimal interference with the integrity of the roof surface.
- The energy yield from the array is higher than mounting the modules on a facade.
- Being high up, shade from trees and neighboring buildings is usually avoided – also true for roof integrated systems.
- Theft is unlikely to be a problem.
- They are easy to disassemble and take down.

Figure 3.17: Small roof mounting structures are very common. They are relatively easy to install and economic

The advantages of roof integrated module mounting systems

- Roof integrated systems are the most aesthetically pleasing and some, such as *solar slates* or *solar tiles,* go well on the roofs of older buildings subject to conservation orders.
- The energy yield from the array is higher than mounting the modules on a facade.
- Being high up, shade from trees and neighboring buildings is usually avoided.
- Theft is unlikely to be a problem.

The advantages of mounting on flat roofs or on the ground
- The modules can be installed at optimum orientation and inclination.
- Good airflow at the back of the modules.
- The installation is very straightforward, speedy and economical.
- Modules can easily be removed for repairs and maintenance (though theft of modules might be made easier).
- Mounting modules on the ground is even more straightforward, speedy and economical, and the load bearing capacities of a roof does not need to be taken into consideration. However, it may be difficult to avoid shading in built-up areas.

Figure 3.18: PV modules in horizontal/landscape layout on a flat roof. The wind forces acting on them are less than would be the case if the modules were vertical/portrait (Photo: B. Breid)

The advantages of facade mounting
- Modules on the facades of prestigious buildings can be an alternative to expensive cladding – what the cladding would have cost can be deducted from the overall cost of the installation.
- Module facades are aesthetically very impressive and very visible.
- There is usually no shortage of surface area on which to mount the modules.

However, facade mounted PV arrays have significantly lower yields because of the vertical inclination.

Figure 3.19: In order to achieve sufficient module cooling, PV facades should, if possible, have air flow behind the array (Source: www.schott.com)

3.6 The site survey

The site survey is an essential part of system design because it is then that most of the information needed to design the system is collected. The first thing that needs to be done is to assess the suitability of the site – usually a roof in the case of grid-tied PV systems. The essential information required is:

- the orientation of the roof
- the angle of inclination of the roof
- the surface area of the roof
- any possible sources of shading.

Ideally, the roof should be shade-free. Often it is not possible to be sure that this will be the case just from looking at the roof. Where there is any uncertainty, a proper shading analysis needs to be made (see 3.10 *The problem of shading* and 3.10.2 *Shade analysis aids* in particular). A professional solar contractor should always visit the site with a *Solar Pathfinder* or equivalent device. In the case of ground mounting, it needs to be ascertained that there is sufficient area and also that the array will be shade-free.

Other information required is:

- the structure and type of roof
- the height of the roof – for scaffolding requirements
- any plumbing or other vents, skylights or other objects to work around
- possible routes for cables
- possible inverter locations
- the intakes from the grid, meters, fuse box, single- or three-phase supply
- the building electrical system grounding/earthing system
- the presence of any lightning protection.

A sketch of the building layout with dimensions should be made and these days, with digital cameras, taking photos of the roof and building is easy. The orientation of the building should be noted. A good compass can be used to do this, but the difference between true north and magnetic north needs to be taken into account where this is significant.

The information collected during the site survey will be used to prepare a cost estimate, so it should be detailed enough to do this. It will also be used to estimate the output of the array. While the situation at the time of the site visit will form the basis of the contract between the customer and the installer, however, the actual installation might take place several months later. If the situation at the site has changed e.g. internal renovations which make a previous planned cable route no longer usable or erection of a building or microwave tower nearby which casts a shadow, this could well increase costs as well as affect the performance of the system. This should be made clear to the customer and written into any contracts. If trees have to be cut back or overhead cables relocated in order to avoid shading the array or any other works required, this should be discussed and recorded. In cases where there are grants or other forms of public funding involved, the information needed for any official application form also needs to be collected. It is important to be informed exactly what details are required for this. There is a sample site survey form for grid-tied PV systems in 7.4 *Site survey form – grid-tied PV systems.*

The site survey is also an opportunity to speak to the customer. It is unlikely that the customer will know a lot about PV systems and this should be borne in mind. (See also 1.5.3 *Initial consultancies with customers* and 1.5.7 *Opportunities for marketing other renewable energy technologies.*)

3.7 Lightning and surge/over-voltage protection

Lightning and surge/over-voltage protection in PV systems is a complex subject. It is beyond the scope of this book to do anything more than provide some background information. It is necessary to refer to national codes and regulations. Some of these will stipulate lightning protection in some parts of a country under certain conditions, but not in others, and they will also specify how it is to be achieved. The design and installation of lightning protection requires specialist expertise. Incorrectly installed lightning protection systems can actually increase the risk of danger to persons and damage to equipment. Lightning protection is also linked to grounding/earthing and codes and regulations regarding grounds differ considerably from one country to another (see 4.5 *Grounding/earthing).* Utilities may also need to be consulted. Stand-alone systems may have different requirements (see 6.6.3 *Lightning protection in stand-alone PV systems).*

In general, the installation of a PV array on a building does not increase the risk of a lightning strike if the height of the building has not been increased or only marginally so. However, on flat roofs, the modules will increase the height of the roof. Ground mounted arrays also need protection. Each installation needs to be approached on a case by case basis. The more expensive an installation is, the greater the need to protect it. But even the best lightning protection system is not a replacement for adequate insurance. And some insurance companies may insist on lightning protection.

If there is a lightning protection system already on the building, the PV installation will generally need to be connected to it. However, the existing lightning protection system might be old and not compliant with the current code and need to be upgraded if changes and additions are made to it. It might also be damaged.

The long shadow cast by a lightning conductor can cause significantly reduced array output. This can particularly be a problem with arrays on flat roofs and with ground mounted arrays. So if one is being installed, its position relative to the array needs to be taken into consideration and lightning protection system designers and installers need to be aware of this. The problem may be overcome by placing the lightning conductor behind the array (relative to the sun) or possibly the use of a shorter lightning conductor.

PV systems can be damaged by direct lightning strikes and by nearby lightning strikes, which induce voltages, currents and magnetic fields in the installation, sometimes even destroying it completely. In general, the effects of lighting strikes – from direct strikes to those occurring at distance, can be categorized as follows (however, note that different countries may have different categorization systems):

- direct strikes – if there is no lightning protection system, the lightning will flow over the installation and more or less destroy it completely – structural damage and fire can also result
- indirect strikes – in this case, the lightning currents flows through the installation cables and utility cables – serious damage is likely
- near strikes – (< 500 m) – magnetic fields induce voltage surges in cables which can cause damage
- distant strikes (> 1000 m) – mainly capacitive effects, usually not so dangerous.

Figure 3.20: Direct lightning strike on a module – cell surface delaminated and shattered glass (Source: Dehn + Söhne & Co. KG)

Figure 3.21: Voltage surge in an inverter – burnt-out capacitor, cable connections and electronic components (Source: Dehn + Söhne & Co. KG)

There are two complimentary types of systems which will protect a building and the PV installation on it: external and internal systems. An external system deals with a direct strike by using one or more lightning conductors which attract the lightning and allow it to flow to ground/earth via protective conductors. Internal lightning protection reduces the risk from the voltage surges induced in the installation and refers to all the measures taken against the effects of the strike and associated electrical and magnetic fields on the metal work and components of the installation being protected.

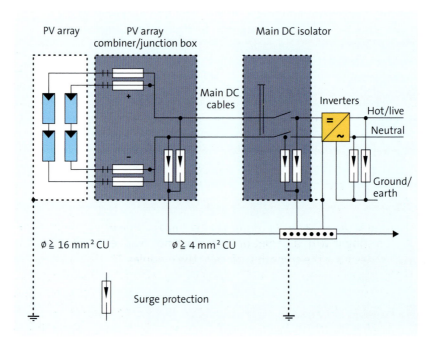

Figure 3.22: Wiring diagram of a grid-tied PV system with surge protection – lightning protection not included

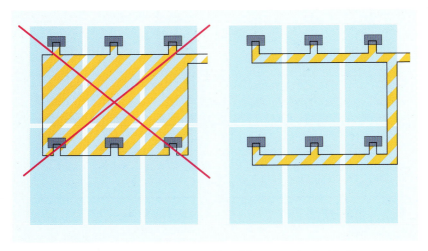

Figure 3.23: Surge protection measure – DC cables of the same string bundled together to avoid loops in which voltage surges can be induced

3.8 Fire hazards and protection measures

A PV array exposed to more or less any level of solar radiation can produce a hazardous voltage. It will do so under conditions of diffuse radiation, i.e. in cloudy conditions and at dawn and dusk. For this reason, a fire on a PV array during the hours of daylight needs to be treated as an electrical fire, and persons extinguishing it need to be aware of this and use the means appropriate to dealing with such a fire, i.e. fire extinguishing foams suitable for applying to electrical fires. However, at night, when an array is exposed only to moonlight or street lighting, arrays do not produce a significant voltage.

There is no technical means of de-energizing a PV array during the hours of daylight. During a fire, covering the modules completely is not practicable. Covering it with a carpet of foam does not work either because the foam slides from the smooth surface of the modules. The Munich Fire Brigade School in Germany did some experiments in which they attempted to black out a solar modules tilted at a range of angles using different types of fire extinguishing foams. They found that they all slid off and that the voltage of the modules was back to its original value in about five minutes.

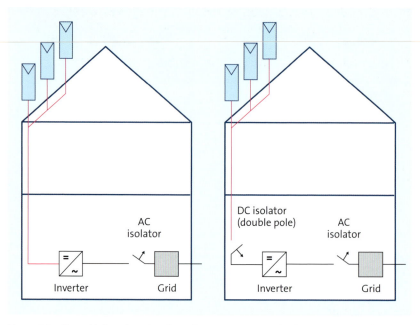

Figure 3.24: The cable in red coming from the PV array cannot be de-energized whether there is a DC disconnect/isolator (right) or whether there is not (left). It will produce a voltage in conditions of daylight (Source: Munich Fire Brigade School)

3.9 Designing grid-tied PV systems – some examples

The design and sizing of a grid-tied PV system depends essentially on the following:
- the size of the roof, its orientation and angle of inclination
- freedom from shade
- technical specification of the modules and inverter
- the aesthetic requirements of the owner/builder/architect
- available finance – private, grants, tax incentives, preferential tariffs
- geographical location, longitude and latitude
- yearly temperature range at the site.

The extent of a suitable roof surface (or other) area and the available budget will usually set the upper limit on system size. An array of 1kWp requires the following surface area, depending on module type:

Type of cell	Monocrystalline	Polycrystalline	Thin film
Surface area required for 1 kWp	6–9 m^2	7–10 m^2	15–20 m^2

The system design process can be divided into the following stages:

1. Initial estimation of system size based on available finance and suitable roof area.
2. Selection of solar modules and compatible inverter(s).
3. Working out the optimum module-inverter configuration.
4. Drawing up a components list.
5. Estimation of yield based on a simulation using solar insolation data for the location.
6. Dealing with any additional aspects – such as lightning protection.

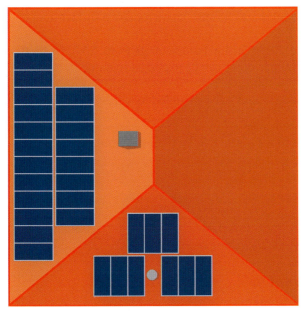

Figure 3.25: The size of the array depends essentially on the available shade-free surface area of the roof and its orientation. Here is a possible module layout on a roof with surfaces facing south and east, with possible shade due to a ventilation pipe

The following worked examples will go through these stages. The first example is done in more detail in order to clarify the individual steps. The other examples are gone through more briefly and go into detail only on those points which differ essentially from the first. For approaches to estimating the outputs of the systems concerned see 2.5 *Estimating the output of PV systems, array angles and orientation* and 2.6 *Sizing and design software*.

Complete systems can also be purchased. In that case, the installer supplies the data and the supplier does the design and sizing. However, the installer will still have to come up with a sensible layout of the modules on the roof. Some suppliers offer computer simulation programs which can help with this.

3.9.1 Example 1: Suburban family home, no shading, 5kWp

Step 1 – Initial estimation of system size

The circumstances
- A suburban home owner is prepared to invest a maximum of € 35,000 in a grid-tied PV system.
- The available roof surface area is 51m²
  (length L_D = 8.5 m and width W_D = 6.0 m).
- The orientation of the roof is due south, and it is inclined at an angle of 45°.
- The roof is completely free of shade.
- Estimated net installed cost is € 5.500/kWp.
- The home owner wants the highest possible output from the array, so monocrystalline modules are chosen – these require a surface area of 9 m²/kWp.
- Temperatures at the site in Central Europe range from -10 °C to 40 °C, module operating temperatures can be expected to range from -10 °C to 70 °C.
- Longitude and latitude would also be needed for estimation of yield.

Initial estimate of system size
Divide the amount of money available by the estimated cost of the system per kWp:

$$\frac{€\ 35{,}000}{€\ 5{,}500/kWp} = 6.36\ kWp$$

and divide the available roof surface area by the roof surface area required by the module type, monocrystalline in this case:

$$\frac{51\ m^2}{9\ m^2/kWp} = 5.67\ kWp$$

So the largest array we can get on the roof 5.67 kWp, which is within the budget.

Step 2 – Deciding on the initial number of modules needed

In making an initial estimate of how many modules are required, the following data is needed:
- peak power of the array (kWp)
- dimensions of roof surface area available (length and width)
- module dimensions
- string sizing to work with a particular inverter (see later steps).

Module selection

To get the process started, a module which might be suitable is selected. In this case, it is the SP 165-M 24 V which has a peak power of 165 Wp (a fictional module). Its dimensions are L_M = 1.61 m × W_M = 0.81 m (which is about 8 m²/kWp). If it turns out that this module is unsuitable, another will have to be selected and the process gone through again.

In order to do an initial estimate of the number of modules required, first divide the initial estimation of array peak power arrived at in Step 1 (5.67 kWp) by the peak power of the chosen module (165 Wp).

$$\frac{\text{Array peak wattage (Wp)}}{\text{Module peak wattage (Wp)}} = \text{Initial estimate of number of modules required}$$

$$\frac{5,670 \text{ Wp}}{165 \text{ Wp}} = 34.4 \text{ modules}$$

This initial estimate gives us 34 modules (34 × 165 Wp = 5.61 kWp).

Surface area requirements and possible module layout configurations

Now one needs to check if 34 of these modules will fit on the roof.

Modules laid out in landscape format :

$$\frac{\text{Roof length } L_D = 8.5 \text{ m}}{\text{module length } L_M = 1.61 \text{ m}} = 5.27 \text{ m}$$

$$\frac{\text{Roof width } B_D = 6.0 \text{ m}}{\text{module width } W_M = 0.81 \text{ m}} = 7.41 \text{ m}$$

This gives a maximum of 5 × 7 = 35 modules (7 rows of 5 modules) which can be laid out in landscape format.

Modules laid out in portrait format :

$$\frac{\text{Roof length } L_D = 8.5 \text{ m}}{\text{module width } W_M = 0.81 \text{ m}} = 10.49 \text{ m}$$

$$\frac{\text{Roof width } B_D = 6.0 \text{ m}}{\text{module } L_M = 1.61 \text{ m}} = 3.73 \text{ m}$$

This gives a maximum of 10 × 3 = 30 modules (3 rows of 10 modules) which can be laid out in portrait format.

The number of modules chosen, 34, can be retained but the they must be laid out in landscape format. The manufacturer of the mounting structure will specify distances between the modules and distance from the edge of the roof. These should be taken into consideration.

Step 3 – Checking the module voltages

Now the voltage values (high, low, normal operating) of the modules must be determined. The highest voltages are achieved in winter when the modules are cold and, in many places, the lowest in summer when the modules are warm. The information needed from the data sheet for this is:
- voltage V_{MPP} and current I_{MPP} at module maximum power point (MPP)
- open circuit voltage V_{OC} at low temperatures, i.e. at -10 °C.

The values for V_{MPP} und I_{MPP} as well as V_{OC} under Standard Test Conditions (at 25 °C) will be on be the data sheet. The open circuit voltage V_{OC} at both extremes of temperature range at the site must be calculated using the voltage temperature coefficient given in the data sheet.

Module specifications on the data sheet

MPP-voltage	V_{MPP}	(at 25 °C)	= 35.35 V
MPP-current	I_{MPP}	(at 25 °C)	= 4.67 A
Open circuit voltage	V_{OC}	(at 25 °C)	= 43.24 V
Short circuit current	I_{SC}	(at 25 °C)	= 5.10 A
Voltage temperature coefficient	T_C	(V_{OC})	= -168.636 mV / °C
Current temperature coefficient	T_C	(I_{SC})	= 2.0 mA / °C
Power coefficient	T_C	$(P_{NOMINAL})$	= -0.420 % / °C

Calculating what the voltage will be at -10 °C and at 70 °C

The yearly temperature range at the location is between 10 °C and 70 °C. So the deviation from Standard Test Conditions (25 °C) for -10 °C is 35 degrees C and for 70 °C is 45 degrees C. The voltage temperature coefficient T_C (V_{OC}) = -168.636 mV / °C, which means that, for every °C the temperature of the module drops below 25 °C, the module voltage will rise by 168.636 mV, or 0.168636 volts (1000 mV = 1.0 volt). And for every °C the module temperature rises above 25 °C, the module voltage will drop by 0.168636 V. Thus adding / subtracting the voltage changes caused by temperature changes to initial voltage values, we obtain:

V_{OC} (at -10 °C)   = 43.24 V + 35 (0.168636 V)   = 49.14 V

V_{MPP} (at -10 °C)   = 35.35 V + 35 (0.168636 V)   = 41.25 V and

V_{MPP} (at +70 °C) = 35.35 V - 45 (0.168636 V)   = 27.76 V

The highest voltage V_{OC} will be at -10 °C , i.e. 49.14 V. And the voltage range at MPP will be between 27.76 V and 41.25 V.

 Many computer simulation programs will calculate the voltage range automatically. Also, most grid-tie inverter manufacturers have sizing programs available on their websites that calculate voltage ranges automatically.

Step 4 – Inverter selection

How many inverters?

In installations of up to 5 or 6 kWp and where the roof surface has a uniform orientation and inclination and is free of shade, the use of a single inverter usually makes sense. In bigger installations the use of multiple inverters can reduce the risk of the whole array going off line, which could happen if a fault occurred in a single central inverter.

The inverter power rating

The power rating of the inverter is determined by the peak power of the array. The peak power of the array is for Standard Test Conditions (STC: $1.000 \, W/m^2$, 25 °C, AM = 1.5) which in practice rarely occur. For this reason, the size of the inverter can usually be about 5 % to 10 % lower than the peak power rating of the array, but the maximum input currents and voltage of the inverter should never be exceeded. National codes also need to be referred to in sizing the inverter as requirements can differ from country to country.

Inverter power rating = 0.90 ... 0.95 × PV array peak power, or

PV array peak power = 1.05 ... 1.10 × inverter power rating.

In Step 2 the peak power of the array was determined to be 5.61 kWp (34 Module × 165 Wp), so an inverter of 5.61 × (0.90 ... 0.95) = 5.05 ... 5.33 kW would be ideal. If the array is to be installed in a less than optimal position – for example an East or West facing roof, or at a steep inclination as on a facade, the inverter can be undersized by more than 10 %. However, experience is needed when doing this and codes need to be referred to (see also 3.2.6 *Inverter – PV array compatibility, inverter size and location* and 4.4.5 *Grid-tied inverter installation and wiring*).

Selecting inverter type

For shade-free arrays of this size central inverters are recommended. The modules are connected in strings which are then connected in parallel in the PV combiner box and then connected to the inverter (see 3.2.2 *Inverter design concepts – central inverters, single- and multi-string inverters and module inverters*). For partially shaded arrays see 3.10.3 *Solutions to shading*. Other inverter features that will influence the choice of inverter are its suitability for installation outside the building, the temperature range it is designed to operate in, indicators and data logging features.

The selection of an inverter with a nominal rating of between 5.05 kWp and 5.33 kWp would be correct, but as we shall see in Step 6, the final number of modules decided upon will be 30 × 165 Wp modules. The peak power of the array will thus be 4.95 kWp. So the rating of the inverter will be between 4,950 Wp × 0,90 = 4,455 Wp and 4,950 Wp × 0,95 = 4,703 Wp. An inverter with a rating of 4.50 kW can be selected.

The data sheet for this inverter gives the following specifications:

Maximum PV Array Power $P_{PV\,MAX}$	= 6.00 kW
DC Nominal Power $P_{DC\,NOMINAL}$	= 4.50 kW
Minimum Peak Power Tracking Voltage $V_{PV\,lower}$	= 125 V
Maximum Peak Power Tracking Voltage $V_{PV\,upper}$	= 750 V
Maximum DC Input Voltage $V_{DC\,MAX}$	= 750 V
DC Nominal Current $I_{DC\,NOMINAL}$	= 9.30 A
Maximum DC Current $I_{DC\,MAX}$	= 22.50 A

Step 5 – Checking voltage limits and module configuration

The aim of this step is to decide on the number of modules in a string. The string voltage needs to be within both the upper and lower limit of the inverter MPP voltage range, i.e. the voltage range within which the inverter will track the MPP of the string. The open circuit voltage of the string also needs to be checked to ensure that it is below the maximum inverter input voltage. The maximum MPP voltage of the modules occurs at −10°C because the voltage of crystalline cells rises as the temperature rises. The minimum MPP voltage will occur at +70 °C (see *Step 3*).

How many of the modules can be connected in series?

The modules are usually series-connected in strings, one for each inverter DC input terminal. The input DC voltage range of the inverter will determine the number of modules to be connected together in each string, as follows:

$$\text{Maximum number of modules} = \frac{V_{PV\,upper}}{V_{MPP}\,(\text{at -10 °C})} = \frac{750\text{ V}}{41.25\text{ V}} = 18.2$$

$$\text{Minimum number of modules} = \frac{V_{PV\,lower}}{V_{MPP}\,(\text{at +70 °C})} = \frac{125\text{ V}}{27.76\text{ V}} = 4.5$$

So, in order to stay within the voltage range at which the inverter will track the MPP of the array, the number of modules in each string must not be fewer than 5 and not be more than 18.

The maximum voltage at the inverter input will occur at -10 °C during open circuit operation (a cold winter day, the sun suddenly appears from behind the clouds). The number of modules in strings must be chosen so that under no circumstances does the string voltage go above the DC voltage input range of the inverter. If it did, the inverter could be damaged.

$$\text{Maximum number of modules} = \frac{V_{DC\,MAX}}{V_{OC}\,(\text{at −10 °C})} = \frac{750\text{ V}}{49.14\text{ V}} = 15.3$$

So the original figure of 18 modules in series has been reduced to 15 modules in order to keep below the maximum DC voltage input of the inverter.

Step 6 – Array configuration – inverter compatibility

Now it is necessary to check if the total number of modules originally decided upon can be divided into strings of equal numbers. The strings need to be of equal numbers because a central inverter has been selected. However, a more expensive multi-string inverter (which can take up to 3 strings) could take strings of different numbers of modules.

$$\frac{\text{Planned number of modules}}{\text{Number of Modules per string}} = \text{Number of strings}$$

A configuration of 34 modules cannot be configured into 15 module strings. Besides, the layout of 34 modules on the available roof surface area might not look very elegant. At this stage it is necessary to discuss with the customer if an array of 30 modules is acceptable. This could consist of 2 strings of 15 or 3 strings of 10 modules. It might even be decided to go for a multi-string inverter to which 34 modules could be connected – 2 strings of 11 modules and 1 string of 12 modules. Or even 35 modules – 1 string of 11 modules and 2 strings of 12 modules.

The design and sizing process involves going through the available options and coming up with the optimal solution. After going through the previous step, it is possible that the total number of modules, the array peak power, the type of modules or the inverter might have to be modified again. This may need to be done several times and may involve doing some of the calculations again, as has been done below.

Now it is necessary to check that the voltages of these strings are within the MPP voltage range of the inverter and do not exceed the maximum acceptable inverter input voltage and that the MPP-current of the strings does not exceed the maximum DC current of the inverter.

Configuration A: 10 modules in series, 3 strings in parallel

V_{MPP} (at 70 °C) = 10 × 27.76 V = 277.6 V.
This value is above the lower limit of MPP voltage range $V_{PV\ lower}$ (125 V).
Acceptable.

V_{MPP} (at −10 °C) = 10 × 41.24 V = 412.4 V.
This value is below the upper limit of the MPP voltage range $V_{PV\ UPPER}$ (750 V).
Acceptable.

V_{OC} (at −10 °C) = 10 × 49.14 V = 491.4 V.
This value is below the maximum acceptable inverter input voltage $V_{DC\ MAX}$ (750 V).
Acceptable.

The current at MPP I_{MPP} of a string has a value of 4.67 A. Connecting 3 strings in parallel will give 14.01 A. This value is above the inverter DC nominal current $I_{DC\ NOMINAL}$ but still below the maximum DC current $I_{DC\ MAX}$.

Configuration B: 15 modules in series, 2 strings in parallel

V_{MPP} (at +70 °C) = 15 × 27.76 V = 416.4 V.
This value is above the lower limit of MPP voltage range $V_{PV\ LOWER}$ (125 V).
Acceptable.

V_{MPP} (at −10 °C) = 15 × 41.24 V = 618.6 V.
This value is below the upper limit of the MPP voltage range $V_{PV\ UPPER}$ (750 V).
Acceptable.

V_{OC} (at −10 °C) = 15 × 49.14 V = 737.1 V.
This value is below the maximum acceptable inverter input voltage $V_{DC\ MAX}$ (750 V).
Acceptable.

The current at MPP I_{MPP} of a string has a value of 4.67 A. Connecting 2 strings in parallel will give 9.34 A. This value is just under the inverter DC nominal current $I_{DC\ NOMINAL}$, and below the maximum DC current $I_{DC\ MAX}$.

Deciding for configuration A or configuration B

In both configurations voltage and current output values are within the respective voltage and current input ranges of the inverter. Either is possible. The more modules connected together in a string the higher the voltage will be (the sum of all the voltages in series). Because higher voltage means lower losses in cables (less voltage drop), configuration B (15 modules) would be preferable to configuration A (10 modules). Connecting the modules in series is also usually quicker and simpler and reduces costs and the possibility of incorrect connections.

Summary

Thus, keeping within the budgetary constraints and the size of the available roof surface, the following configuration emerges:

- PV array peak power – 4.95 kWp
- number of modules – 30 × 165 Wp modules
- configuration – 2 strings of 15 modules
- central inverter with a DC nominal power rating $P_{DC\ NOMINAL}$ of 4.50 kW and maximum PV array power rating $P_{PV\ MAX}$ of 6.00 kW.

Other configurations are possible using other modules and inverters. But bear in mind that the availability of modules and inverters can limit the choices available. The planning process described above does not need to be done manually. Design and sizing software or the inverter manufacturer's web-based sizing programs make the design process a lot easier.

Many manufacturers and system suppliers offer a free design service and have a hot line to give advice during planning and installation. Free software is also available which will quickly give array configuration options and even provide a list of components. It is also a good idea to double check designs with manufacturers and system suppliers.

3.9.2 Example 2: Barn, sloping roof, circa. 15 kWp, 30 kWp

Step 1 – Initial estimation of system size

An estate owner would like to have a grid-tied PV system installed on one of the estate buildings. There is no upper limit on the budget. If possible, the PV array should cover the whole of the roof. The temperature range is the same as for the previous example.

The roof surface area is 130 m² (length L_D = 13.0 m, width B_D = 10.0 m). The orientation of the roof is south-south-east (-20°), and its inclination is 30° from the horizontal. The array size can be estimated:

$$\text{PV array peak power} = \frac{130 \text{ m}^2}{(8 \text{ m}^2/\text{kWp})} = 16.25 \text{ kWp approx.}$$

Step 2 – Deciding on the initial number of modules needed

The module initially selected, the SP VPM M 175 has a peak power of 175Wp. Its dimensions are L_M = 1.58 m × W_M = 0.85 m (which is about 7.7m²/kWp).

$$\frac{16,250 \text{ Wp}}{175 \text{ Wp}} = 92.8$$

which gives an initial figure of 92 modules, and an array peak power of 92 × 175 Wp = 16.1 kWp.

Now one needs to check if 92 of these modules will fit on the roof.

Modules laid out in landscape format:

$$\frac{\text{Roof length } L_D = 13.0 \text{ m}}{\text{module length } L_M = 1.58 \text{ m}} = 8.23$$

$$\frac{\text{Roof width } B_D = 10.0 \text{ m}}{\text{module width } W_M = 0.85 \text{ m}} = 11.76$$

This gives a maximum of 8 × 11 = 88 modules (11 rows of 8 modules) which can be laid out in landscape format.

Modules laid out in portrait format:

$$\frac{\text{Roof length } L_D = 13.0 \text{ m}}{\text{module width } W_M = 0.85 \text{ m}} = 15.29$$

$$\frac{\text{Roof width } B_D = 10.0 \text{ m}}{\text{module } L_M = 1.58 \text{ m}} = 6.33$$

This gives a maximum of 15 × 6 = 90 modules (6 rows of 15 modules) which can be laid out in portrait format. The previously estimated number of modules must be reduced from 92 to 90. And these will have to be laid out in portrait format.

Step 3 – Checking the module voltages

The information on the module data sheet is:

MPP-voltage V_{MPP} (at 25 °C)	= 35.4 V
MPP-current I_{MPP} (at 25 °C)	= 4.95 A
Open circuit voltage V_{oc} (at 25 °C)	= 44.4 V
Short circuit current I_{sc} (at 25 °C)	= 5.40 A
Voltage temperature coefficient T_c (V_{oc})	= -156 mV/ °C
Current temperature coefficient T_c (I_{sc})	= 2.9 mA/ °C
Power co-efficient T_c ($P_{NOMINAL}$)	= -0.485 %/ °C

Using the voltage temperature coefficients, we can now determine the voltage at the extremes of the temperature range:

V_{oc} (at -10 °C)	= 44.4 V + 35 (0.156 V)	= 49.86 V
V_{MPP} (at -10 °C)	= 35.4 V + 35 (0.156 V)	= 40.86 V
V_{MPP} (at +70 °C)	= 35.4 V - 45 (0.156 V)	= 28.38 V

Step 4 – Inverter selection

Inverter nominal power = 0.90 ... 0.95 × array peak power, so the size range will be from 15,750 Wp × 0.90 = 14,175 Wp to 15,750 Wp × 0.95 = 14,963 Wp. An single inverter of this size is not available, so several inverters must be used. Thus 3 inverters, each with a nominal rating of 4.80 kW, are selected. Each has a maximum PV array power rating of 6.00 kW. The nominal rating of the 3 inverters together is 14.40 kW and their total maximum PV array power rating is 18.00 kW.

The data sheet for the inverters gives the following specifications:

Maximum PV Array Power $P_{PV\ MAX}$	= 6.00 kW
DC Nominal Power $P_{DC\ NOMINAL}$	= 4.80 kW
Minimum Peak Power Tracking Voltage $V_{PV\ lower}$	= 350 V
Maximum Peak Power Tracking Voltage $V_{PV\ upper}$	= 650 V
Maximum DC Input Voltage $V_{DC\ MAX}$	= 750 V
DC Nominal Current $I_{DC\ NOMINAL}$	= 12.00 A
Maximum DC Current $I_{DC\ MAX}$	= 15.00 A

Step 5 – Checking voltage limits and module configuration

$$\text{Maximum number of modules} = \frac{V_{PV\ UPPER}}{V_{MPP}\ (at\ -10\ °C)} = \frac{650\ V}{40.86\ V} = 15.9$$

$$\text{Minimum number of modules} = \frac{V_{PV\ LOWER}}{V_{MPP}\ (at\ +70\ °C)} = \frac{350\ V}{28.38\ V} = 12.3$$

So, in order to stay within the voltage range at which the inverter will track the MPP of the array, the number of modules in each string must not be less than 12 and not be more than 15 and,

$$\text{Maximum number of modules} = \frac{V_{DC\ MAX}}{V_{OC}\ (at\ -10\ °C)} = \frac{750\ V}{49.86\ V} = 15\ \text{modules}$$

which means that that number of modules will also not exceed the input voltage of the inverter.

Step 6 – Array configuration – inverter compatibility

Configuring 90 modules in strings of 15 modules is possible with 6 strings. Using 3 inverters, it would be divided up as follows: 2 strings of 15 modules each per inverter. This configuration has to be checked to see if it is compatible with the MPP tracking voltage range, the maximum input voltage and the maximum input current of the inverter:

V_{MPP} (at +70 °C) = 15 × 28.38 V = 425.7 V.
This value is above the lower limit of MPP voltage range $V_{PV\ LOWER}$ (350 V). Acceptable.

V_{MPP} (at -10 °C) = 15 × 40.86 V = 612.9 V.
This value is below the upper limit of the MPP voltage range $U_{PV\ UPPER}$ (650 V). Acceptable.

V_{OC} (at -10 °C) = 15 × 49.86 V = 747.9 V.
This value is below the maximum acceptable inverter input voltage $V_{DC\ MAX}$ (750 V). Acceptable.

The current at MPP I$_{MPP}$ of a string has a value of 4.95 A. Connecting 2 strings in parallel will give 9.90 A. This value is below the inverter DC nominal current I$_{DC\ NOMINAL}$ and well under the maximum DC current I$_{DC\ MAX}$.

The configuration is within all the voltage and current input ranges of the inverter.

The 15 kWp (approx.) PV array in this example is constructed modularly. More of these *modules*, each one consisting of a 5 kWp (approx.) array of 30 modules, could be installed if there was sufficient roof surface area. For example, 150 of the same modules could be installed with 5 of the same inverters.

If it is not possible to divide the array into equal parts, each with its own inverter, it can be divided into unequal parts. For example, an array of 27 kWp could be divided into 4 × 5 kWp and 2 × 3.5 kWp.

Summary

An array of 15.75 kWp consisting of 90 × 175 Wp modules, configured in 6 strings of 15 modules each, can be installed on the available roof surface area. Using other modules and inverters, other configurations are possible.

3.9.3 Example 3: Warehouse, flat roof, circa. 30 kWp

Step 1 – Initial estimation of system size

A haulage contractor would like to have a PV array installed on a warehouse. There is no upper limit on the budget. If possible, the PV array should cover the whole of the roof. The temperature range is the same as for the previous example.

The roof surface area is approximately 1,000 m² (length L$_D$ = 35.0 m, width B$_D$ = 29.0 m). The orientation of the roof is due south. The roof is flat but the modules will be tilted at an angle of 30° from the horizontal, the optimum angle for a grid-tied system at the latitude of the site.

With the above figures we can do the following initial calculation:

$$\text{PV array peak power} = \frac{1{,}015 \text{ m}^2}{(9 \text{ m}^2/\text{kWp})} = 112.8 \text{ kWp approx.}$$

If it was a sloping roof, it would be possible to have an array in the region of 112.8 kWp, but on a flat roof the rows of modules will have to be spaced to avoid shading in winter. In this case and at this latitude in Central Europe, the rows need to have a space between them which is 3.5 times the height of the rows. This can be worked out geometrically by calculating the length of shadow cast by the sun at around mid-morning in mid-winter, but it is probably easier to consult with suppliers of this type of flat roofing system. As a result, the approximate size of the array will be smaller, as follows:

$$\frac{112.8 \text{ kWp}}{3.5} = 32.3 \text{ kWp}$$

So the peak power of the array will be around 32.3 kWp.

Step 2 – Deciding on the initial number of modules needed
The module initially selected, the SP VPM P 260, has a peak power of 260 Wp. Its dimensions are L_M = 1.61 m × W_M = 1.34 m² (which is about 8.3 m²/kWp).

$$\frac{32,300 \text{ Wp}}{260 \text{ Wp}} = 124.2$$

which gives an initial figure of 124 modules, and an array peak power of 124 × 260 Wp = 32.2 kWp. Now one needs to check if 124 of these modules will fit on the roof.

Modules laid out in landscape format:

$$\frac{\text{Roof length } L_D = 35.0 \text{ m}}{\text{module length } L_M = 1.61 \text{ m}} = 21.74$$

$$\frac{\text{Roof width } B_D = 29.0 \text{ m}}{\text{module width } W_M = 1.34 \text{ m}/3.5} = 6.18$$

This gives a maximum of 6 × 21 = 126 modules (6 rows of 21 modules) which can be laid out in landscape format.

Modules laid out in portrait format:

$$\frac{\text{Roof length } L_D = 35.0 \text{ m}}{\text{module width } W_M = 1.34 \text{ m}} = 26.12$$

$$\frac{\text{Roof width } B_D = 29.0 \text{ m}}{\text{module } L_M = 1.61 \text{ m}/3.5} = 5.15$$

This gives a maximum of 26 × 5 = 130 modules (5 rows of 26 modules) in portrait format. So the original number of 124 modules can be laid out in portrait, and this format is preferable (see 3.4 *PV module mounting structures and systems*).

Step 3 – Checking the module voltages

The information on the module data sheet is:

MPP-voltage V_{MPP} (at 25 °C)	= 57.1 V
MPP-current I_{MPP} (at 25 °C)	= 4.55 A
Open circuit voltage V_{oc} (at 25 °C)	= 70.9 V
Short circuit current I_{sc} (at 25 °C)	= 4.91 A
Voltage temperature coefficient T_c (V_{oc})	= -269 mV/°C
Current temperature coefficient T_c (I_{sc})	= 4.9 mA/°C
Power co-efficient T_c ($P_{NOMINAL}$)	= -0.470 %/°C

Using the voltage temperature coefficients, we can now determine the voltage at the extremes of the temperature range:

V_{oc} (at -10°C) = 70.9 V + 35 (0.269 V) = 80.31 V,

V_{MPP} (at -10°C) = 57.1 V + 35 (0.269 V) = 66.52 V and

V_{MPP} (at +70°C) = 57.1 V − 45 (0.269 V) = 45.00 V

Step 4 – Inverter selection

Inverter nominal power = 0.90 … 0.95 × array peak power, so the size will be from 32,240 Wp × 0.90 = 29,016 Wp to 32,240 Wp × 0.95 = 30,628 Wp. An inverter with a nominal power rating of 27 kW is selected. It has a maximum PV array power rating of 33 kW.

The data sheet for the inverter gives the following specifications:

Maximum PV Array Power $P_{PV\ MAX}$	= 33 kW
DC Nominal Power $P_{DC\ NOMINAL}$	= 27 kW
Minimum Peak Power Tracking Voltage $V_{PV\ BOTTOM}$	= 450 V
Maximum Peak Power Tracking Voltage $V_{PV\ TOP}$	= 800 V
Maximum DC Input Voltage $V_{DC\ MAX}$	= 900 V
DC Nominal Current $I_{DC\ NOMINAL}$	= 50.00 A
Maximum DC Current $I_{DC\ MAX}$	= 60.00 A

Step 5 – Checking voltage limits and module configuration

$$\text{Maximum number of modules} = \frac{V_{PV\ UPPER}}{V_{MPP}\ (\text{at -10 °C})} = \frac{800\ V}{66.52\ V} = 12.0$$

$$\text{Maximum number of modules} = \frac{V_{PV\ LOWER}}{V_{MPP}\ (\text{at +70 °C})} = \frac{450\ V}{45.00\ V} = 10.0$$

In order to stay within the voltage range at which the inverter will track the MPP of the array, the number of modules in each string must not be fewer than 10 and not be more than 12.

$$\text{Maximum number of modules} = \frac{V_{DC\,MAX}}{V_{OC}\,(\text{at -10 °C})} = \frac{900\,\text{V}}{80.31\,\text{V}} = 11.2\,\text{modules}$$

which means that that number must be reduced from 12 to 11 in order not to exceed the maximum DC input voltage of the inverter.

Step 6 – Array configuration – inverter compatibility

Configuring 124 modules in strings of 11 modules is not possible. However, 121 modules with 11 strings of 11 modules is. But this would lead to an inverter loading of 116 %. So it is decided to go for 110 modules – 10 strings with 11 modules in each. This gives an inverter loading of 106 %. This configuration has to be checked to see if it is compatible with the MPP tracking voltage range, the maximum voltage and the maximum current of the inverter:

V_{MPP} (at +70°C) = 11 × 45.00 V = 495.0 V.
This value is above the lower limit of MPP voltage range $V_{PV\,LOWER}$ (450 V). Acceptable.

V_{MPP} (at -10 °C) = 11 × 66.52 V = 731.7 V.
This value is below the upper limit of the MPP voltage range $U_{PV\,UPPER}$ (800 V). Acceptable.

V_{OC} (at -10°C) = 11 × 80.31 V = 883.4 V.
This value is below the maximum acceptable inverter input voltage $V_{DC\,MAX}$ (900 V). Acceptable.

The current at MPP I_{MPP} of a string has a value of 4.55 A. Connecting 10 strings in parallel will give 45.5 A.. This value is just below the inverter DC nominal current $I_{DC\,NOMINAL}$ and well under the maximum DC current $I_{DC\,MAX}$.

The configuration is within all the voltage and current input ranges of the inverter.

Summary

An array of 28.60 kWp consisting of 110 × 260 Wp modules, configured in 10 strings of 11 modules each and connected to a central inverter is an acceptable option.

3.10 The problem of shading

The shading of modules should be avoided altogether. Recurring and even occasional shadows on a PV array will drastically reduce its output. This chapter explains why shading is such a problem, how to avoid it and, where it is inevitable, what can be done to mitigate its effects. Ideally, no shade should be cast on a south-facing array between 10.00 and 15.00 hours on the day of the winter solstice (21st December), the day when the sun is lowest in the sky. All calculations relating to shade analysis should be made on this basis.

Figure 3.26: If at all possible, shading should be avoided. It is one of the main reasons for low outputs.

If some array shading is inevitable, good design can mitigate its effect. The issue is complicated by the fact that shadows move with the sun, but their movement – like the movement of the sun – can be predicted. However, this is particularly difficult when the shadows are cast by several objects at different heights, of different dimensions and at different distances away, all the more reason to avoid shading in the first place. But, if some shading is inevitable, it needs to be taken into consideration when planning both the layout of the modules and the string configuration. However, the design of PV arrays subject to shade should only be carried out by experienced designers.

Why shading is so important – the garden hose effect

When a solar cell is shaded, it can no longer produce current. It then behaves like a blocking diode and current cannot flow in the other cells in the series either. This is the so-called *garden hose effect* – no water comes out of a hose, even if it is blocked in only one place. And in addition to this, because the shaded cell acts as a blocking diode, it is subject to a voltage which is the sum of the voltages of the other cells connected in the series, which is higher than the breakdown voltage of the diode. If the diode breaks down, the cell can get extremely hot and can cause permanent damage to the module. This is the so-called *hot spot* effect.

When solar cells are connected in series, the current produced by that series of cells is determined by the lowest current produced by any cell in the series and, since the cell current is determined by the intensity of solar radiation falling on it, the effect of shading on even a single cell is to reduce the current produced by all the cells in the series. The same is true for modules – the lowest current produced by a module in the series will determine the current output of the entire string of modules. Thus even the shading of one cell reduces the output of the modules which reduces the output of the string which reduces the output of the entire array.

3.10.1 Types of shading

Occasional shading

Occasional shading is caused by leaves, bird droppings and other dirt. If the array is cleaned regularly by the rain, most of this type of dirt will be washed off. An inclination of 15° is sufficient and, at a steeper angle, the water flowing over the array will clean it even more effectively. The edges of modules or a mounting structure can be a hindrance to this and over time strips of dirt above the lower edges of module frames can accumulate and cause permanent shading. In climatic zones where there are long dry periods, a layer of dust can accumulate on the modules and they will have to be washed – by hand using a wet sponge if they are accessible or simply hosed down if they are out of reach. Snow usually simply slips off modules – generally, the more northerly the installation the steeper the tilt of the array in any case.

Recurring shading

Recurring shading can be caused by parts of the building on which or near which an array is mounted or by items on the roof or in the vicinity of the array, such as chimney stacks, projecting parts of the roof and facade, unusual roof structures, dormer windows, satellite dishes, trees etc. Overhead power and telecom cables can also be a source of shade, as can lightning conductors, especially if they are too high and too thick or too near the modules or are too numerous. These types of objects cast a thin shadow but it is sharp and moves over the array. Neighboring buildings and trees can also cause shade, as can the elevated horizon in a deep valley.

Figure 3.27: The chimney stack on this roof is shading one of the modules. The array has not been positioned to avoid it (Photo: Matthias Belz)

3.10.2 Shade analysis aids

Shade analysis aids make it much easier to predict where recurring shading will occur. The analysis is made on location. It enables the designer to get a silhouette of the landscape from the point of view of the PV array and identify objects which are potential sources of shade. There is now software which enables photos from digital cameras to be used to analyze array shading in computerized simulations.

Figure 3.28/3.29: Shading can be predicted using a solar path indicator – needs to be for correct latitude; or the Solar Pathfinder – works at all latitudes, with the correct paper latitude chart inserted. (Thomas Seltmann/www.solarpathfinder.com)

3.10.3 Solutions to shading
There are basically two types of solutions to shading: either remove the source of the shade or, if it is inevitable and its effects more or less acceptable, design and install the system so that its effect will be minimized.

Removing source of the shading
Some sources of shade can be removed:
- trees can be moved and replanted, as long as they are small
- tree branches can be cut back – but this needs to be done regularly
- satellite dishes and antennae/aerials can be moved – but make sure that when birds perch on them their droppings will not fall directly onto the array instead
- overhead cables can be rerouted or buried.

Bypass diodes
Small shadows and dirt such as leaves and bird droppings can significantly reduce the output of crystalline cells. Bypass diodes mitigate the effect of this somewhat and avoid hot-spot damage. They are connected in parallel to strings of cells (see illustration). The diode allows the current to bypass the affected cell. In commercially available modules, a maximum of 24 cells will be protected by one bypass diode. The diodes are usually installed in the module junction box. Diodes to protect whole strings are often found in the PV array combiner box. Bypass diodes also make it possible to physically position the modules so that the effects of shade is mitigated (see below).

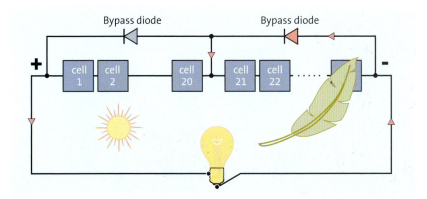

Figure 3.30: The reduced output and possibility of damage to cells and modules caused by shade can be mitigated by the use bypass diodes. The diode short circuits the affected area and allows the current to bypass it

Inverters and shading
The use of a central inverter is problematic if there is partial shading of an array because it will have to process the output of both the shaded and the shade-free modules, effectively causing *mismatching*. The inverter will find a MPP which is less than optimal. The use of several single-string

inverters or a multi-string inverter is recommended instead when partial shading is unavoidable.

Optimum arrangement of modules

Where it is not possible to avoid shade, its effect can be mitigated by physically positioning the modules in a more appropriate arrangement, by appropriate string configuration and use of several or multi-string inverters.

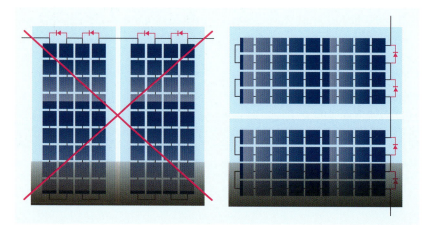

Figure 3.31: On the left only the lower half of the bottom module is shaded – output is reduced by 25 %. On the right all cell strings are affected – output is reduced by 100 %. Note the diodes

Figure 3.32: The modules on this house are arranged around the satellite dish / aerial in order to avoid shade (Photo: Dürschner).

Figure 3.33: The roof of this industrial premises has numerous protruding ventilations pipes. The area around this pipe has been covered with inexpensive dummy modules. The uniform appearance of the array has been preserved and the shadows cast by the ventilation pipes avoided (Photo: Dürschner).

Module string configuration and shade

Here are two examples of how module configuration can be used to deal with shading.

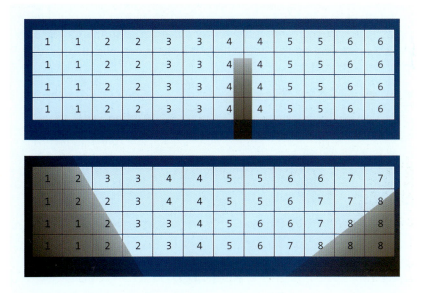

Figure 3.34: Optimal array configuration in the case of vertical shading (above) and shading from the side (below). The top array has strings of 8 modules and the lower array has strings of 6 modules – the strings are numbered

EXAMPLE 1 – vertical shadow moving across a roof: an industrial chimney south of the array is casting a thin vertical shadow which moves from left to right over the course of the day. By configuring the array in 6 strings of 8 modules each (see photo – top half) only one string is shaded at a time.

EXAMPLE 2 – shading from the side – mornings and evenings: in the mornings a triangular shadow is being cast by a neighboring building to the south on to the left corner of the array, and in the evenings a neighboring building on the east side is casting a similar shadow on to the right of the array (see photo – lower half). By configuring the array in 8 strings of 6 modules each only a few strings are affected.

Module mounting structures and shade

When modules are being mounted on free-standing structures in rows, either on flat roofs or on the ground, attention needs to be paid to the following:
- sufficient space needs to be left between the rows of modules to avoid them shading each other
- if the structure is ground mounted, it needs to be high enough so as not to be shaded by growing plants – or the grass needs to be cut regularly
- in heavy rains, the bottom of the module can be splashed with mud – and become dirty – this needs to be avoided
- where snow is an issue, the modules need to be installed high enough off the roof so that when the snow slips off them it does not gather below and shade them that way.

Rows of modules installed in landscape format need less distance between them than modules installed in portrait format because they are not as high, but there will be fewer modules per row. And for the same reason, rows of modules with shallow inclinations will also need less space between them.

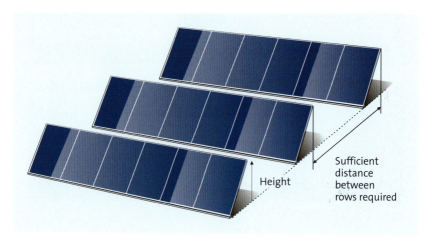

Figure 3.35: The distance between rows of modules installed in rows on a flat roof or on the ground should be such that the rows of modules do not cast shadows on those behind them.

If there is limited space on a flat roof or area of ground, it may be possible to reduce the module inclination in order to have more modules. While the output of the individual modules will be reduced slightly, the overall array output will be increased because there are more modules. However, caution is called for, depending on latitude.

Sizing software and shade

Predicting the output of PV arrays which are subject to shade is difficult even for someone who is very familiar with simulation software. Inputting data regarding array shading into the simulation programs is time consuming and subject to error. And doing a plausibility check on the result is even more difficult again.

3.11 Preparing cost estimates and quotes

The most expensive item in any grid-tied PV system will be the PV modules. Other costs include the mounting structure, the inverter, the PV combiner box, the DC disconnect / isolator and the DC cables. The installation itself will take time and scaffolding or hydraulic lifts are often required. There are also the costs of site visits, transport, planning, on-going consultation with the customer and administrative costs. A solar business will also have the usual business overheads and marketing costs, all of which have to be worked into the price (see also *1.4.6 The economic viability of PV* and *2.4.12 The cost of modules*).

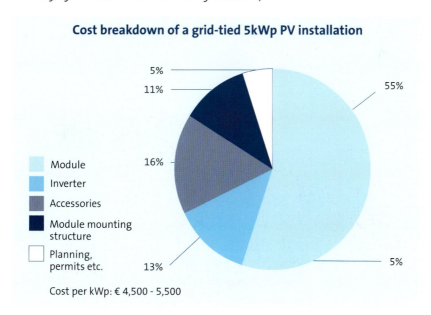

Figure 3.36: Percentage costs of system components in a 5 KWp grid-tied PV system (Source: Solarenergieförderverein Bayern e.V., Munich, Germany)

While, conditions which influence how long a job will take and cost vary considerably from installation to installation and from country to country, some general points can be made:

- larger modules are generally easier to install than smaller ones
- larger installations cost less in terms of € per kWp than smaller ones
- some roofs will be more difficult to work on than others
- arrays installed on several surfaces cost more than those on a single uniform surface
- the module mounting structure can influence the cost
- the height of the roof must be taken into consideration
- not having all the necessary components to hand and non-standard constructions will add to the cost
- lightning and voltage surge protection will add to cost, but it may not be necessary
- experience pays, as a rule of thumb the first installation will take twice as long as the 4th, and four times as long as the 20th.

4 Grid-tied PV Systems – Installation and Commissioning

This chapter is a general guide to the installation of grid-tied PV systems. The previous chapter which describes these components and their functioning should be read first and in conjunction with this chapter. National codes and guidelines and manufacturer's manuals always need to be referred to. A list of relevant national electrical codes and guidelines is given in 7.1 *Installation codes and guidelines*.

 All electrical work carried out on grid-tied PV systems should be carried out by suitably qualified persons who are aware of national code requirements with regard to the installation itself and safe working practices. The work involves higher voltages than are found on regular building distribution wiring, working with DC electricity rather than the more familiar AC and other particular hazards. Installers should also know how to deal with an electrical accident should one occur.

 During installation work, grid-tied PV systems should be considered as building sites and all the safety precautions normally associated with building sites should be taken. Working on roofs is particularly hazardous and national guidelines and safe working practices should be adhered to.

4.1 Overview

The installation and commissioning of grid-tied PV systems is regulated by different laws and electrical codes in different countries. In some places these laws and regulations will be very comprehensive and in others not, generally depending on the state of development of the industry. Similarly, it is easier dealing with utilities who have previous experience. Anyone installing grid-tied PV needs to familiarize him-/herself with the local legislative and regulatory framework and comply with it. All installation instructions and manuals should be read carefully.

The actual procedures to be followed will probably look something like this:
1. Obtaining all necessary permits.
2. Installation of the system.
3. Inspection and testing of the system.
4. Final permission to connect.
5. Actual connection to the grid and system commissioning.
6. System handover to the customer.

Installers need to be familiar with relevant codes, legislation and utility requirements covering the following issues:

- regulations regarding who is permitted to install grid-tied PV systems
- qualifications required for personnel carrying out work
- permits required from utilities for grid-connection
- necessary building permits
- specifications of approved equipment, particularly of the inverter
- procedures to be followed when connecting to the utility grid
- inspection and testing
- certification requirements
- net-metering requirements
- conditions to be met in order to obtain public funding/tax relief/ preferential tariffs.

4.2 Preliminary paperwork

Before work on the actual physical installation can begin, some permits will usually be required. These include permission to connect to the grid, building permits and net-metering agreements. Note that the grid utility from whom one gets permission to connect, may not necessarily be the utility to whom the electricity produced by the array is sold. The application for permission to connect to the grid will very often be via a standardized form available from the utility or used nationally. All relevant system specifications will have to be provided. This may include extra documentation certifying that the inverter is of an approved type. The same information or similar information will also be necessary if the customer is applying for public funding/tax relief/preferential tariffs.

4.3 System installation

4.3.1 Who is allowed to install a grid-tied PV system?

Who is allowed to install grid-tied PV systems will differ from country to country and even between different jurisdictions within a country. The installer will usually need to be registered with one or more bodies and meet the conditions required by them. This can cover such items as the qualifications of the personnel doing the work, insurance and previous experience. For the customer to qualify for public funding/tax relief/preferential tariffs, the installer usually needs to be registered with the body responsible for administering these incentives.

4.3.2 Working on roofs – safety issues

Working on roof is hazardous. National health and safety legislation and guidelines should be strictly complied with. No work should be carried out on any PV arrays if there is any thunder or lightning in the area.

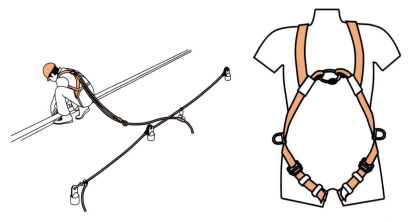

Figure 4.1: Appropriate safety measures should be taken by all persons working on roofs

4.3.3 Electrical work on PV arrays – particular hazards

Working with electricity is hazardous and there are additional hazards associated with working with PV arrays. These can be summed up as follows:

- PV arrays always produces a voltage when exposed to light – this cannot be switched off on the DC side of a grid-tied installation
- PV arrays will continue to produce a voltage even when there is a fault or short circuit
- PV arrays are a limited current source – fuses and circuit breakers do not operate in the way they do in normal AC mains power circuits
- PV arrays are often configured to produce relatively high and hazardous DC voltages
- opening or separating contacts while a PV array is under load (feeding current onto grid) can lead to arcing which does not self-extinguish
- danger of falls from roofs and other structures due to electric shock.

 When a PV array is exposed to light, it cannot be de-energized in the same way as, for example, a diesel generator can. In one sense, it can be *switched off* using the DC disconnect / isolator, but the array and all the associated wiring is still hot / live on the supply side of the DC disconnect / isolator (see diagram in 3.8 *Fire hazards and protection measures*).

 A PV array is also a limited current source. The short circuit current is not much higher than the maximum normal operating current. When there is a fault in normal wiring connected to the grid, the short circuit current that rushes in from the grid is so high that it more or less instantaneously blows the fuses or opens the circuit breakers. This does not happen with a PV array. The short circuit current that occurs under fault conditions is simply too low. A short circuit in the array wiring will also not stop the array producing voltages. In the event of electric shock from contact with exposed hot / live parts, the current that flows under such conditions will not be interrupted by the string fuses or other circuit breakers, so it can last a very long time.

PV arrays are often configured to produce relatively high and hazardous DC voltages – much higher than AC distribution circuit voltages (240/400VAC in Europe and 120/230VAC in North America) – than electricians are used to working with. From the point of view of electric shock, voltages above 120 VDC and 50 VAC are considered hazardous. Even very low currents – 200 mA DC and 50 mA AC – for more than 200 milliseconds (mS) can cause a fatal heart attack. PV modules produce voltages of between 20 VDC and 100VDC. When modules are connected in series the voltage is the sum of the voltage of the modules in the series. Thus, a shock hazard exists even when two modules are connected together in series. Even at lower voltages touching both the positive and negative conductors at the same time should be avoided.

The DC voltages found in PV arrays can cause arcing which is not self-extinguishing. It can occur at voltages above 50 VDC. Arcing can occur when a contact is opened/broken while the array is under load, such as when a terminal is disconnected, a plug unplugged or probes of an ammeter removed. The result can be burns, flashover and electric shock. These arcs can weld cable ends together. There are modules on the market which have an open circuit voltage which exceeds 50 VDC. If the positive and negative leads of such a module are connected together under condition of full sunlight, an arc will occur when they are separated again. Disconnections on the DC side of the installation – in junction boxes or of plugs/sockets – should never be made while the array is connected to the inverter. Open the DC disconnect/isolator first. When the array is connected to the grid through the inverter, there is a flow of current. Breaking this current can lead to dangerous arcing. Arcing can also occur in DC switches, disconnects and isolators which may not be rated for DC load breaking.

While national codes need to be referred to for comprehensive details and specifics of protective measures to be taken when working in grid-tied PV systems, the following can serve as a summary:

- all the usual safety precautions that are taken when working on electrical circuits should be adhered to, taking into consideration the additional hazards associated with working with PV arrays (see table)
- single core and double insulated cables should be used, if available
- enclosures and accessories should be double insulated
- the PV array should be disconnected from the inverter by having the DC disconnect/isolator in the open/off position when any work is being done on the DC side of the installation.
- hot/live working should be avoided wherever possible
- all hot/live parts such as cable ends which need to be temporarily exposed should be carefully insulated using insulating tape
- any damaged cables should be dealt with immediately and safely by taping them temporarily with insulating tape, eventually repairing or replacing them
- electrical hazards can be almost eliminated by disconnecting one series connector in the middle of each PV string.

General electrical safety measures	PV array particularities
Isolate the supply from source of electrical energy (open disconnects / isolators, remove fuses) when doing work	PV array cannot be switched off when exposed to light, but can be disconnected from inverter by opening DC disconnect / isolator
Ensure against accidentally reconnection to supply (put fuses in pocket, lock disconnects / isolators in open position)	
Check that there is no voltage on conductors etc.	When exposed to light, it is not possible to make PV array cables voltage-free except by disconnecting them from the modules
Make sure hot / live parts cannot be touched (placing out of reach, barriers, fixing covers, insulation tape etc.)	

All the usual electrical safety precautions need to be taken when working with PV arrays and electricians need to be aware of additional potential hazards

4.3.4 Installing the modules – general guidelines

Modules can be installed on roofs, on free-standing structures, on roofs or on ground-mounted structures. There are so many module mounting structures and systems on the market that it would not be useful to go into any of them in any detail here. Good systems will either be largely self-explanatory or have a good explanatory manual. Leading manufacturers will also have a telephone advice line. This can be very useful for obtainig advice generally and clearing up ambiguous points in instruction manuals (see also 3.4 *PV module mounting structures and systems* and 3.5 *Where to install: roof or facade?*).

Module mounting structures on roofs

There are nearly as many module mounting systems and structures as there are types of roofs. And there is a great variety of roofs – even within localities. Roofs are designed to withstand local climatic conditions, often built using local materials and using local building techniques. Roofs in Alaska will be very different from roofs in Florida. Visiting PV installations in the locality to see what local practice is can be well worthwhile. That said, it is possible to give some general guidelines:

- some manufacturers and suppliers will help with the first installation or will offer short courses – it is worthwhile taking advantage of any offers like this
- good installation instructions for module mounting structure to go on roofs will describe and show very clearly how the structure is to be fixed to the roof and the modules to the structure.

- engaging the services of a roofer or builder with experience of working on roofs is a good idea when first installing roof mounted systems and/or when working on a difficult roof, at least to install the fixing points onto which the actual module structure or rack is to be fixed; on some types of roofs this may always be necessary.
- when working with other companies or trades people, it is important to be clear about who is responsible for what and who provides warranties for what
- installing an array on an old roof or one in a state of disrepair is not recommended, the PV array will survive much longer than the roof, so the roof should be renovated beforehand – renovating (or trying to) the roof with the array on it several years down the line will be significantly more expensive
- modules should be installed with sufficient gaps between them and behind them to facilitate airflow and cooling
- cables and accessories should be suitable for external use
- it is a good idea to read manuals and instructions for roof mounting structures even before the structure is purchased – any potential problems can be identified well in advance
- installers need to be aware of relevant building regulations and what specifications mounting structures need to comply with
- the services of a structural engineer may be necessary at times.

Figure 4.2: Installing modules on a tiled roof. The DC cables are being fed through the opening in a ventilation tile. It is important that cables are not subject to abrasion (Photo: Seltmann).

Systems with modules integrated into roof structures

Roof integrated mounting structures involve removing the roof surface (or part of it) and replacing it with modules/laminates/*solar tiles*/*solar slates*. There is a limited range of products on the market. Most of the general points made regarding module mounting structures on roofs are applicable. Sufficient air flow under/behind the modules is of primary importance. If there is not enough airflow, the temperature of the modules can rise by more than 20 °C in comparison to well ventilated ones. In hot conditions this will reduce array output by 10 % to 30 %. It is essential that the finished job is entirely watertight

Figure 4.3: These solar modules/tiles, being installed in Adelaide, Australia, will be part of the roof structure. Standing on PV modules needs to be done with great care (Source: www.pvsolartiles.com by PV Solar Energy Pty Ltd Australia)

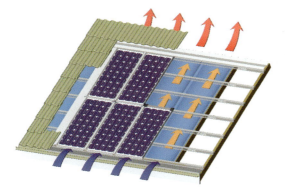

Figure 4.4: It is important to ensure there is an airflow behind the modules/tiles in roof integrated systems, especially in hot climates. Or consider using amorphous silicon (Source: www.pvsolartiles.com by PV Solar Energy Pty Ltd Australia)

Mounting structures on flat roofs and free-standing structures
- Making holes in the roof surface should be avoided. If unavoidable, it should be done by a specialist.
- Cables should be fed over the edge of the roof and through the side walls of the building, rather than through the roof.
- The amount of weight needed to keep free-standing structures in place depends on the strength of winds the structure will be exposed to.
- The load-bearing capacity of the roof will set an upper limit on the amount of weight that can be put on it and it will determine how they are distributed over the roof.
- On a flat roof, protective matting needs to be placed under the support points and the weights of the structure.
- Cables should be installed in conduit or trunking to protect them against UV and weathering. At the very least they should be laid in the shade.

4.4 DC cable installation – general guidelines

The guidelines given above in 4.3.3 *Electrical work on PV arrays – particular hazards* should be followed. National codes will specify the specifics but in general:
- cables with double insulated and polarized (reverse polarity protected) DC connectors should be used
- disconnections of any kind (including removal of string fuses) should not be made when the array is under load because of the danger of arcing
- cables being fed into a building should be looped to prevent ingress of rain drops
- cables should not be left hanging free or laid on concrete or other surfaces, they should be tied up and out of the way with UV-resistant cable ties (usually black) or enclosed in solid or flexible conduit.
- large loops should be avoided. cables should be laid parallel to each other and close together – 3.7 *Lightning and surge/over-voltage protection*
- cables should be as far as possible from lightning protection equipment and not cross over lightning conductors and associated cables
- DC cables should not be laid in spaces in which there is a particular fire danger
- cables from the module strings should follow the shortest route to PV array combiner boxes
- all DC cables in the building should be clearly identifiable as such, particularly when they are installed near to or together with the building's AC and other cables
- before cables are energized, all continuity, insulation and other tests specified in national electrical codes should be carried out.

Avoid laying DC cables through spaces in which there are flammable materials or with a potentially combustible atmosphere, such as garages and workshops. Where it is unavoidable, relevant electrical codes should be referred to.

4.4.1 DC wiring – modules and strings

The main points here are:
- the type and size of cable specified in national electrical codes should be used – the main features of these are outlined in 3.3.1 *Cables*
- care should be taken that the modules are connected up correctly; in grid-tied systems usually in series-connected strings in which the positive pole of one module is connected to the negative pole of the next (see 3.2.1 *Grid-tied inverters and module configurations* for configuration diagrams)
- all cables should be given string numbers – this will aid any later fault finding and repair.
- cables on roofs should not interfere with the flow of rain water from the roof
- refer to 4.3.3 *Electrical work on PV arrays – particular hazards.*

Modules with double insulated and polarized (reverse polarity protected) DC connectors should be used in installations with string inverters. There will be no PV array combiner box, so connectors may need to be unplugged from modules to ensure that cables are not energized when connecting them into the inverter.

Testing

In addition to the continuity and insulation tests required on wiring before it is energized, the following tests should also be carried out on each string:
- verify DC polarity, i.e. that the conductors identified by color coding as positive are in fact positive and those identified as negative are in fact negative
- check the open circuit voltage V_{oc} of the string; this will be close to the sum of the open circuit voltages of all the modules in the string
- check the short circuit current I_{sc} of each string; under clear sky conditions, this will be about 80–95 % of the I_{sc} on the data sheet (measured at 1,000 W/m² = full sun) and about 10 % of that in cloudy and overcast conditions; make sure that the instrument has a sufficient current rating

These tests should verify that the strings have been configured correctly and with the correct number of modules. National codes and guidelines need to be referred to for full testing requirements (see also 4.8 *System commissioning).*

Minimal instrument requirements are:
- continuity and insulation tester as specified in national codes
- digital multimeter, capable of measuring up to 10 ADC.

Verifying DC Polarity

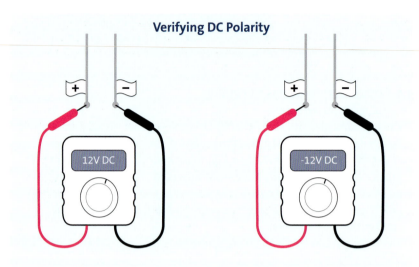

Figure 4.5: DC polarity is verified using a digital millimeter set at the DC voltage range. The cables have been tagged + and – and the test is to verify if they have been tagged correctly. If polarity is incorrect, the instrument will give a voltage reading with a minus (-) sign in front of it. Make sure that the instrument leads are in the correct instrument sockets. Tests for V_{OC} and I_{SC} are made in a similar manner. Instruments need to be adequately rated for expected voltages and currents. The cables on the right have been correctly tagged, those on the left incorrectly

4.4.2 PV array combiner box and terminating the strings

Installations with central inverters will need a PV array combiner box for each inverter. The main points here are:

- the PV array combiner box should be able to house the required number of string cables
- if located near the inverter it can also incorporate the main DC disconnect / isolator
- some codes may require that, if it needs to be fixed onto wood or any other potentially combustible structure, it should be mounted on a fire-proof plate
- it should be located as close as possible to the modules in order to keep the string cable run as short as possible and enable optimum functioning of surge protection equipment
- if installed externally, it should be rated for external use and appropriate fixings and sealants should be used to prevent moisture ingress
- the cables should be numbered to indicate which string they belong to and color coded
- positive, negative and grounding / earthing conductors should be physically separated from each other by non-conducting plates or junction boxes.

The integral protection class of the PV array combiner box should be appropriate to the situation in which it is being installed, i.e. if installed externally, it should be rated for external use. A short circuit in the DC combiner box due to ingress of moisture is extremely dangerous and a serious fire hazard.

Touching the exposed contacts in the DC junction is extremely dangerous. The voltages and associated currents are very high. Short circuits can create arcs which are not self-extinguished. These can cause serious injury, burns and fires.

If the modules are connected together in strings using plug/socket connectors, one of them on each string should be left open while working on the PV array combiner box, the DC disconnect/isolator or the inverter. They should be plugged together only when the system is ready to be commissioned. This means that hot/live working is largely avoided.

4.4.3 DC wiring from PV array combiner box to inverter

The DC cables from the PV array combiner box to the inverter (via the main DC disconnect/isolator) are referred to as the *main DC cable(s)* below. The exact specification of these cables and how they are to be installed will be outlined in national codes, but in general:

- the main DC cables should be rated for at least the full short circuit current of the PV array and the voltage drop should usually not be greater than 2%
- codes will specify what type of cable is required – armored cables may be acceptable
- connection of the main DC cables to the terminals in the PV array combiner box should not be made when the input terminals in the DC are hot/live; this can be achieved by opening the DC disconnect/isolator in the PV array combiner box, if it has one; or by unplugging or otherwise disconnecting the DC cables from the array strings; the DC disconnect/isolator should be kept in the open position until the system is ready to be commissioned.
- having the positive and negative conductors in separate cables provides a higher level of protection against accidental short circuit, even if this is not specified in the national code; otherwise mechanical damage or overheating (occurring under fault conditions or during a fire) can lead to short circuit/arcing; a distance of 50 mm between them should be sufficient
- they need to be clearly and correctly color coded
- the main DC cables in particular should be protected from mechanical damage by humans and animals.

Unlike the wiring in the rest of the building, the main DC cables on the PV array side of the DC disconnect/isolator are not de-energized by opening the DC disconnect/isolator. They need to be clearly labeled, so they cannot be confused with the other wiring in the building.

4.4.4 The main DC disconnect/isolator

The main DC disconnect (sometimes called the *main DC isolator* or *main DC switch)* needs to be installed adjacent to the inverter. Its function is to enable isolation of the PV array from the inverter so that work can be performed on the inverter in voltage free conditions and to enable disconnection of the PV array from the grid. There are PV array combiner boxes and even inverters which have integrated DC disconnects/isolators. Bear in mind that, in the case of an inverter, if an integrated DC disconnect/isolator has to be removed and one without a DC disconnect/isolator installed in its place, the installation will be without one. This would not be code compliant.

- The DC disconnect/isolator needs to be rated for code-specified DC current and DC voltage ratings. These values will be based on the open circuit voltage V_{oc} and the short circuit current I_{sc} of the PV array. A disconnect/isolator that can break the circuit under load is usually recommended, if not specified in codes. A sign affixed at/or near the disconnect/isolator will also usually be required.
- If the inverter is near the PV array combiner box, the ideal solution is to have the DC disconnect/isolator integrated into the PV array combiner box.

4.4.5 Grid-tied inverter installation and wiring

All the documentation that comes with the inverter should be read and kept. Copies should be given to the customer.

Inverter location

- The inverter should be installed as near as possible to the modules in order to minimize DC cable lengths.
- Inverters installed outside should be rated for a temperature range of at least -25 °C to +60 °C, and have the appropriate integral protection (IP) rating.
- Locations with potential sources of moisture (damp cellars, above a washing machine) should be avoided.
- Dusty locations should be avoided.
- Locations which will cause the inverter to be heated in addition to its normal heating during operation should be avoided. The location should also remain cool in summer.
- Inverters are relatively heavy and should only be installed on walls that can bear their weight.

- Do not install the inverter to the outside walls of bedroom or living rooms – some vibrate and make a buzzing noise when they are operating. A 2 kW inverter with a transformer with a 35 dbA buzz even in the hallway of a house can be disturbing.
- Accessibility for future maintenance and repairs needs to be considered. Work platforms and ladders which were available during the original installation work will usually not be available.
- Theft and vandalism can also be an issue if the location is easily accessible to the public.

 During normal operation 3 % to 5 % of inverter nominal power will be converted into heat and at times this will reach 10 %. For a 5 kW inverter, this works out at 150–250 W and on occasion to 500 W, in effect converting the inverter into a small electric heater. Inverters that get too hot in summer and consequently derate themselves are a common cause of reduced array yields.

Warm locations such as the following are to be avoided:
- unconverted spaces under the roof, particular on the south side – these can get very hot in summer
- the walls of chimneys
- locations exposed to direct sunlight, also sun traps outside
- built-in cupboards
- locations without air circulation
- inverters installed one above the other

If the inverter has to be installed in a warm location, it should have fan-assisted cooling.

 Avoid dusty locations such as animal feed stores, warehouses storing building materials, barns or carpentry workshops. Dust can block the ventilation vents of the inverter. It will then become too hot and regulate the output of the array downwards.

 Inverters can weigh about 10 kilos per kW. A 4 kW inverter can weigh 40 kg. Serious thought needs to be given to putting a heavy inverter safely into place – specialist equipment may be needed.

Inverter wiring
- Inverter wiring should be carried out with the inverter disconnected from both the PV array and the grid – with the DC disconnect / isolator on the PV array side in the open / off position and the AC disconnect / isolator on the grid side in the open / off position.
- Any data cables for monitoring and programming via remote monitors or modems should be laid at the same time (see 4.4.6 *Installing remote monitors*).

Inverter – PV array side wiring (DC)

The following tests need to be carried out on the wiring on the DC side (PV array side) of the inverter:

1. Continuity and insulation resistance of the DC wiring from the PV array combiner box to the inverter – this will uncover any short circuits, faults to ground/earth and faulty insulation.

2. DC polarity needs to be verified.

3. Array current (with a clamp meter) and voltage.

A clamp meter should be used when measuring current at this point of the circuit, i.e. in the main DC cable(s). The strings are connected in parallel in the PV array combiner box and the current will be the sum of the currents in the strings. Measuring this current using an ammeter with probes would be hazardous as arcing would inevitably occur when the probes were disconnected.

The voltage and current readings should be compared with what is expected. This is the last opportunity to uncover any faults or incorrect numbers of modules in the array and mistakes in configuring the modules.

Inverters are usually not reverse DC polarity protected. Connecting the positive terminal of the inverter to the negative of the array, and vice versa, can destroy the inverter. This will not be covered in the inverter warranty.

Never attempt to measure the current between the positive and negative terminals when the inverter is in operation i.e. when the inverter is connected to the grid. This can destroy the measuring instrument. The inverter has a capacitor in its input which will be short circuited and give rise to a powerful discharge. This, together with the array current, will overload the instrument.

After the array has been disconnected from the inverter by opening the DC disconnect/isolator, there can be a voltage at the input terminals of the inverter which can last for several minutes. The reason for this is that the capacitor is charged during inverter operation.

Inverter – grid side wiring (AC)

See 4.6 *Connection to the grid and meter location* for a range of possible inverter/meter/building fuse box/grid-connection point configurations. National codes and utilities will usually have quite stringent requirements which need to be followed, but in general:

- the size of the AC cable from the inverter to the point at which it is connected to the grid needs to be rated for the inverter AC output current and protected by a fuse or circuit breaker of the appropriate rating; but, it also needs to be sized for voltage drop, which should be no more than 1 % (if not specified differently) at maximum current
- an AC disconnect/isolator (often lockable) needs to be installed to enable the inverter to be disconnected from the grid for maintenance and repairs – the building fuse box disconnect/isolator may be an acceptable alternative however.

It is only possible to communicate with some inverters via modem/remote monitor during the day when the PV array is operational. Some inverters will take their power supply exclusively from the array, not from the grid and need to be in stand-by or operating mode in order to be communicated with.

4.4.6 Installing remote monitors

Some remote monitors can be installed without any wiring, the data being transmitted by radio wave. Other monitoring systems use *power line carrier communication* which transmits the data to a display over the normal house wiring. However, this can be negatively affected by other electrical equipment. Tests need to be carried out to ensure that there is no interference. If there, it is best to install a separate data cable.

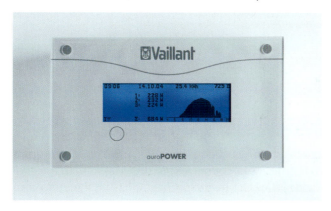

Figure 4.6: Remote monitors enable customers to check system performance from the comfort of their living rooms. Some operate wirelessly (Source: Vaillant)

4.4.7 Fuses and circuit breakers

Fuses and breakers installed on the DC side need to be rated for DC and the appropriate voltage. Also, the PV array is a current limited generator and its short circuit current will only be about 20 % greater than its operating current, which means that fuses and circuit breakers will not give anywhere near the same level of protection that they give in more conventional circuits. National codes should be referred to and see 4.3.3 *Electrical work on PV arrays – particular hazards*.

4.5 Grounding/earthing

The metal frames of modules and the metal mounting structure generally need to be grounded/earthed. As in most electrical installations, the purpose of grounding is to ensure that in fault conditions leading to metal enclosures or structures becoming hot/live, the voltage does not rise significantly above ground/earth potential and to ensure that fuses and circuit breakers disconnect the supply within time limits specified by codes (which, as discussed above, on the DC side of a grid-tied PV installation can be problematic). Usually either the metal components of the PV array will be connected to each other and to the main grounding point or to an electrode which may need to be installed. If there are no metal parts in the array (for example, in a roof integrated system using laminates rather than framed modules), no grounding may be required. Inverters will usually also need to grounded. It is important to bear in mind, that there are not only electrical hazards originating in the PV array itself, but that, under fault conditions the PV modules and array mounting structure can be made hot/live by the grid itself. Some codes may specify the use of ground fault interrupters (GFIs in the USA) (residual current circuit breakers RCCBs or RCDs in the UK), and the type which is to be used. USA codes require GFIs built into the grid-tie inverter. (See also 3.7 *Lightning and surge/over-voltage protection*).

In all decisions regarding grounding/earthing requirements for PV modules frames, PV array mounting structures and grid-tied inverters, national electrical codes need to be consulted and implemented. Regulations regarding grounding/earthing requirements and methods differ significantly from country to country.

4.6 Connection to the grid and meter location

The schematic diagrams in this section illustrate some of the more common configurations for the actual connection of the PV system to the grid. National electrical codes should provide specifications and the utility also needs to be consulted. A schematic of what configuration is to be used should be part of the initial application to the utility for permission to connect to their network. Lockable AC disconnects/isolators are often required. The size of the systems will also have a bearing on the configuration, as will whether the system is single-phase or three-phase or even two-phase.

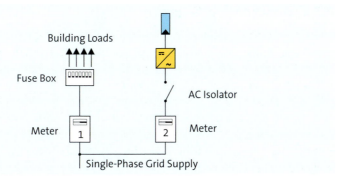

Figure 4.7: In this single-phase configuration, the connection to the grid is made to the grid incoming cable. Meter 1 counts the number of electrical units taken from the grid and meter 2 counts the number of electrical units put onto the grid. Note that there is an AC disconnect/isolator between the inverter and the meter. Suitable for selling all the electricity onto the grid

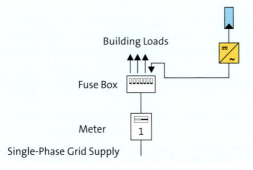

Figure 4.8: In this single-phase configuration the connection to the grid is made via a fuse way in the building main fuse box. The single meter is an import-export meter, no record is made of the number of electrical units produced by the PV array, only of what is fed onto the grid and taken from it. The number of units put onto the grid will be less than the number of units produced by the array because some of the electricity will be used in the building itself i.e. the building's electricity consumption is offset by the electricity produced by the PV array so it does not have to import as much as it otherwise would have. The main switch on the fuse box can be used to isolate the array from the grid. However some national electrical codes insist on a dedicated and lockable AC disconnect/isolator between the inverter and the fuse box

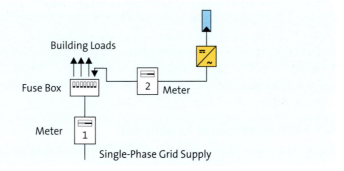

Figure 4.9: In this single-phase configuration the connection the connection to the grid is made via a fuse way in the building main fuse box. Meter 2 counts the number of units produced by the PV array. This enables the output of the array to be easily monitored and the monthly output recorded. Such information on past performance is very important when analyzing problems and trouble shooting

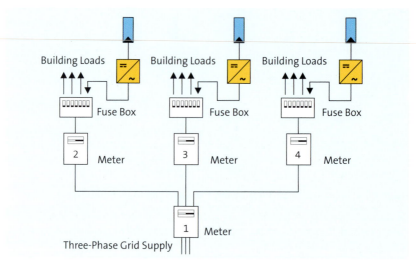

Figure 4.10: In this 3-phase system, 3 small single-phase arrays are connected separately into 3 single-phase fuse boxes. Meter 1 is a 3-phase import-export meter. Meters 2, 3 and 4 are recording the output of the PV arrays. Suitable for offset, import and export

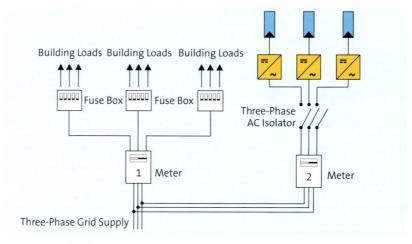

Figure 4.11: In this three-phase system, 3 large PV systems are connected directly to the incoming grid cables. Meter 1 is a 3-phase meter which counts the number of electrical units taken from the grid and Meter 2 is a 3-phase export meter which counts the number of electrical units put onto the grid. Note that there is a 3-phase AC disconnect/isolator between the inverter and the meter. Suitable for selling all the electricity onto the grid

4.7 Final permission to connect

Permission from the utility to connect the PV system to the grid is required. Procedures to be followed will differ from country to country and from utility to utility. In some cases no final permission may be needed after the initial application to connect to the grid has been accepted. Installers need to familiarize themselves with the situation in their locality.

4.8 System commissioning

Commissioning the system, that is switching it on and operating it in parallel with the grid, is usually covered comprehensively in national codes and guidelines. Utilities may also have special requirements. They may insist that a representative inspect and test the installation and/or be present when the installation is commissioned.

The aim of all these inspections, tests and approvals is to ensure that
- the installation is mechanically and electrically sound
- that it does not present a fire or a shock hazard
- that inverter features to prevent *islanding* are operational – see *Inverter grid monitoring and islanding prevention* in *3.2 Grid-tie inverters*.

Final testing will be needed and the results recorded. These can include:
- final visual inspection
- verification that the installation complies with specifications in permits etc.
- accessibility of disconnects/isolators, presence of required signs, drawings
- inverter tests/array tests
- a functional test of inverter safety features
- checking that export-import meter is working properly.

Any hazards, non-compliance with codes, problems found in the AC wiring of a building in which the PV system is being installed should be notified to the customer in writing. This is usually a legal obligation.

What specific electrical tests need to be carried out, how they are to be carried out and with what instruments will be specified by national codes and guidelines. These can include:
- array open circuit voltage V_{oc} (main DC disconnect/isolator in open position)
- string open circuit voltages V_{oc} (main DC disconnect/isolator in open position)
- voltage drop across DC string fuses
- short circuit current I_{sc} of each string (main DC disconnect/isolator in open position – with instrument capable of measuring full string current)
- tests relating to the grounding/earthing of the installation – this may include a ground/earth electrode test for which a ground/earth electrode tester is required
- insulation resistance of the array – care needed/damage can be caused if test is not carried out correctly
- insulation resistance of the main DC cable.

(See also 4.4.1 *DC wiring – modules and strings*).

When carrying out a functional test on the inverter, refer to the inverter manual. Inverters need one to several minutes in order to boot up and operate normally again after being switched off. If the grid goes down 3 times (in reality or in simulation) in close succession, some inverters need up to a quarter of an hour before they begin operating normally again.

4.9 System handover to customer

Handing over the system to the customer is an important part of the job. The installer needs to explain to the customer or whoever will be in charge of the system what the main components of the system are and their functions, as well as the meaning of any indicators etc. A printed list (with system diagram) is useful for this. Any maintenance which is the customer's responsibility needs to be gone through, as well as the details of any maintenance contract. The customer should also be given a complete file of all relevant documentation, including component data sheets, manuals, wiring diagrams, copies of test and inspection certificates, output projections. This should be in a proper folder with installer details – it is likely that it will be shown to other potential customers and is good advertising. It can also be referred to later if any problems occur – it can be difficult to analyze faults in PV systems if one does not know how the system was originally performing. It may also include a form with which the customer keeps a record of the system's output. There may also be official documents to sign.

5 Grid-tied PV – Operation and Maintenance

5.1 Grid-tied PV system operation

Grid-tied PV systems operate fully automatically and will usually do so for many years without any problems. The main thing the owner or the person in charge really has to do is monitor the array output and be able to differentiate between normal operation or if something is wrong so that he or she knows when to get in touch with the installer.

5.2 Monitoring system performance

System owners are interested in system performance, in how much electricity they have sold and / or how much of their normal consumption has been offset by the system, so many are quite willing to keep a record of system performance – at least of total monthly output. These records are important. If any problems develop in a system, knowing how it has being performing makes fault finding a lot easier. Adequate metering and remote monitoring makes this task a lot easier (see 4.6 *Connection to the grid and meter location* and 3.3.5 *Remote monitors*).

5.3 Maintenance and servicing

Grid-tied PV systems are essentially maintenance-free, and they usually operate perfectly for years without any intervention. However, maintenance contracts will ensure better operation over the 20 or more year life of the installation. The level of service required will differ from customer to customer. Some will be more interested in getting involved than others, and some – institutions for examples – will be happy to contract out entirely (see 1.5.4 *Contracts, quotes, estimates and insurance*).

5.3.1 Maintenance – what system owners can do

Installers should ensure that the person responsible for the system is given adequate written instructions and other documents regarding basic system maintenance, and instructed regarding hazards and safety requirements – and owners and operators should also be given sufficient information to differentiate between normal performance and possible under-performance.

Visual inspections

Regular visual inspections should be recommended to look for things like:

- array shading and dirt on the array – pollen in spring, leaves in autumn, snow in winter, tree growth, bird droppings
- dirt gathering near the edges of modules
- storm damage to modules or support structure
- loose cables, cables damaged by rodents, chaffing and abrasion.

System owners and building maintenance personnel need to be aware of the danger of electric shock that can occur if cables or components are damaged and that these should never be touched. Flashover can occur even at close proximity to an exposed fault. The repair and replacement of damaged cables and components should only be carried out by qualified persons.

Cleaning the array

Cleaning the array is usually not necessary. The rain will usually do the job except in very dry or dusty regions. Small quantities of dust or pollen for short periods of time will have no significant effect on the array output. However, longer lasting dirt such as that which may gather above the edges of modules frames or bird droppings needs to be dealt with. Cleaning can be part of a servicing contract – done on a yearly or seasonal basis or according to need. However, if it is to be carried out by the system owner or building maintenance personnel, they need to be aware of all the dangers associated with working on roofs and of the safety precautions that need to be taken. Solvents should never be used, water with a small amount of soap is sufficient.

Snow on the array

In many places snow will not be a problem at all. In fact, snow has less of an effect on the annual array output than is generally assumed. It usually falls at times of the year when output is low anyway (see table). But where there is significant snow over longer periods of time, clearing it from the array needs to be considered, particularly on large and/or flat arrays.

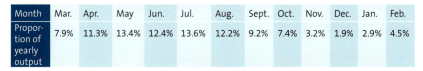

Month	Mar.	Apr.	May	Jun.	Jul.	Aug.	Sept.	Oct.	Nov.	Dec.	Jan.	Feb.
Proportion of yearly output	7.9%	11.3%	13.4%	12.4%	13.6%	12.2%	9.2%	7.4%	3.2%	1.9%	2.9%	4.5%

Output of PV array in Central Europe in monthly percentages of total output. Snow on the array for a few days in January does not make much difference to the yearly yield

5.3.2 Maintenance – what the PV specialist can do

A maintenance contract with the installer means that the system owner can be sure he or she will get the most out of the system. Most electrical codes specify that electrical installations be tested every few years depending on the type of installation. The PV system is part of a building's electrical installation and will normally be encompassed within this. However, most electricians will not be familiar with them. In addition, national codes and/or utilities might insist on periodic inspection and testing, particularly of the inverter safety features. Installers need to be aware of what these requirements are and to inform customers. Legally required periodic inspection and testing can be part of a maintenance contract.

Yearly maintenance visits should, at least, include:
- checking the operation of the inverter via indicators. LEDs, etc,
- cleaning the inverter ventilation vents
- functional test of inverter safety and islanding prevention features
- visual inspection of the array and visible cables
- inspection of module mounting structure frame
- checking the output of string voltages and currents
- documenting any deficiencies.

Additional services that can be offered include:
- monitoring, recording and evaluating system performance
- cleaning the array
- cutting back vegetation causing shade.

 When any maintenance or servicing is being done on the array, the main DC disconnect/isolator should be in the open position (see *4.3.3 Electrical work on PV arrays – particular hazards* and Chapter 4 generally).

5.4 Possible malfunctions in grid-tied PV systems

5.4.1 System producing less than expected
Although PV technology is robust and reliable, problems do occur. If a system is producing less electricity than expected, it could be due to:
- shade from trees, other buildings, overhead cables, aerials
- incorrect inverter size
- regulation problems/defective inverter
- module output in reality lower than that specified
- faults in the DC wiring
- faulty operation of inverter safety features
- defective modules
- defective (module) bypass diodes
- problems caused by the utility grid.

Problems originating on the utility grid cannot be rectified directly by the installer. It is a complex area and their analysis requires a high level of specialist knowledge.

5.4.2 Problems originating on the utility grid

In Germany, where there are more grid-tied systems than anywhere else, experience has shown that the grid itself can be the cause of many of the problems associated with malfunctioning grid-tied PV systems. And in Germany the quality of the grid is very high. The basics of how inverters interact with the grid is discussed in *Inverter grid-monitoring and islanding prevention* in 3.2 *Grid-tied inverters*. These issues are becoming more important because of the increased sensitivity of inverters. A characteristic of these problems is that they are intermittent and last only for a short time, which makes them difficult to identify.

The most common causes of grid-related problems are:
- fluctuating grid voltage
- deviations from nominal grid frequency
- fluctuations in grid impedance.

Many of these grid-related problems are impossible to identify without the help of the utility. Some can be rectified by having the operating parameters of the inverter readjusted by the manufacturer but others will need the help of the utility itself to address. But there is no need for installers to be discouraged by this; most of these issues can be sorted out at the planning stage of an installation and through discussion with the utility.

Fluctuating grid voltages

Problems with grid voltage can be due to:
- voltage too high
- grid voltage too low
- deformed grid voltage (defective or deformed sine wave).

Too high (or too low) a grid voltage

Inverters will switch themselves off if the grid voltage goes above a certain level. Grid-tied PV installations themselves increase the voltage of the grid – they must operate at a slightly higher voltage than the grid in order to feed current onto it. If there is already another PV installation close by, it is possible – for example at midday – it might raise the grid voltage over the acceptable threshold and cause the inverter of a neighboring PV system to automatically switch itself off. If this is the case, the installation of another PV system may depend on some work being done on the grid to rectify the problem. On the other hand, the installation of a new PV system close to an existing and perfectly functioning system may cause the existing system to switch off during periods of good solar isolation.

Problematic higher voltages can also be caused by large inductive loads in the vicinity, such as the large motors of a saw mill or a pumping station or electrical equipment in factories and on industrial estates. Problems such as this can only be addressed by the grid utility itself. In an ideal situation, the grid operators will be informed at the planning stage and will make an assessment of the suitability of the local grid for the number of PV systems to be connected to it. If the grid at that point is unstable or would be overloaded, they can then take appropriate measures, usually by the installation of cables with a greater cross-sectional area. In some cases, adjusting the voltage output of the local transformer will suffice. Either of these measures could involve extra costs. Also, many utilities will have little or no experience with the connection of PV systems. Low grid voltages can also be a problem.

EXAMPLE: An inverter has been switching itself off for short periods of time 50 or 60 times a day. This has been registered in the data logger but there was no indication of what the source of the problem could be. A fluctuation in the grid voltage was not indicated. The system installer informed the manufacturer of the inverter who requested the utility to look into it. It emerged that there was a large water pumping station nearby which was inducing short (millisecond duration) voltage spikes. The inverter data logger was not picking this up because the durations of the voltage spikes were too short. However, the inverter itself was detecting it and disconnecting itself from the grid.

The operating parameters of many inverters can be adjusted so that they can operate reasonably well connected to a less than perfect grid. This is something that needs to be discussed with the inverter manufacturer at the planning stage of an installation.

Deformed grid voltage

EXAMPLE: The inverter is switching itself off several times a day without apparent reason. The inverter manufacturer contacted the utility and the cause was identified: a pig fattening farm. The infra-red lamps are being dimmed 50% by using phase dimming/waveform chopping. This was causing a quasi-DC voltage on the grid which was distorting the shape of the grid sine wave and causing the inverter to switch off.

Note: Grid-tied inverters themselves tend to stabilize the grid because of their *clean* sine waves, in that they impose clean sine waves on any deformed or defective ones. The question of inverters causing harmonic distortion on the grid has been researched. No significant effect has been detected.

Grid frequency variations

Inverters are set to operate in a specific frequency range. Distortion can occur. For example, older wind turbines which are not operating to specified technical standards can affect grid frequency.

Grid impedance variations

Grid-tied inverters will also only operate within a certain grid impedance range and are sensitive to impedance fluctuations. For example, an inverter may have the following grid impedance parameters: maximum Z_{AC} = 2.5 Ω and maximum change ΔZ_{AC} = 0.5 ohm within 0.2 seconds. If the impedance fluctuation ΔZ_{AC} in 0.2 seconds is more than 0.5 Ω, the inverter will switch off. The impedance ranges of inverters can be adjusted if they are too sensitive.

5.5 Fault finding and trouble shooting guide

Problems in grid-tied PV systems are not always easily initially identifiable. Finding out what is actually wrong is the first task. If the performance of the system has been regularly monitored and the records are available, this task is a lot easier. Comparisons with comparable installations in the same locality can also be helpful. In general, system malfunctions and under-performance can be sorted into 3 categories:

1. No system output during the day.

2. Lower system output than expected or in comparison with similar locally-situated system.

3. The output of the system is lower than it used to be.

5.5.1 Malfunction category 1: No system output during the day

Unless the sky is very cloudy or it is raining or the array is covered in snow, the PV system should be generating electricity and this should be going onto the grid or consumed by the loads in the building. It is only on relatively dark days or in the early morning or late evening that the array voltage can be insufficient to do this. At night inverters usually switch themselves off completely.

1. Fault indication	2. Fault indication	Fault / possible cause / solution
Meter/s show no output. Inverter showing no input voltage from the array and no injection onto grid	No DC voltage at the inverter input	Too dark, not enough light. Come back at a better time when there is enough sunlight. If not =>
		Main DC disconnect / isolator in open position? Defective disconnect / isolator? Check voltage at disconnect / isolator input. If not =>
		String fuses blown (lightning strike)? If not =>
		Excess voltage suppressor has short-circuited the array to earth. Check excess voltage suppressor. If not =>
		Open or short circuit in the array? Damaged cables or modules? Visual inspection required. Open PV array combiner box and test strings.
	There is a DC voltage at the inverter input but the inverter indicators are not showing anything	Too dark, not enough light. Come back at a better time when there is enough sunlight. If not =>
		Defective inverter. Contact the manufacturer
	Inverter indicates DC input voltage during the day but nothing is being put onto the grid	Utility grid black-out? Brown-out? Try turning on a light. The inverter should operate again when the grid comes back on. If not =>
		Blown fuses, activated circuit breakers and ground fault interrupters on the AC side between inverter and grid? The main utility fuse? Check these. If not =>
		The inverter has detected a fault in the array and shut down. Check any fault indicators. Test strings individually in the PV array combiner box. Possibly isolate the string which is causing the inverter to shut down by disconnecting one string at a time until string with fault is identified? If not =>
		The inverter has detected a grid fault or grid operating outside design parameters for the inverter causing the inverter to shut down. Check inverter indicators. Inverter should automatically start up again when problem clears. Contact utility if it is reoccurring frequently. (See 5.4.2 *Problems originating on the utility grid*)

5.5.2 Malfunction category 2: System output lower than expected

This category deals with situations in which the output is lower than one would expect from the system specifications or in comparison with a similar system in the same locality. However, when comparing the performance of different systems care needs to be taken. Different orientation, horizon shading, different module types and the site conditions can lead to significant variations in output and performance.

ATTENTION: When comparing the performance of different installations, always compare several with each other (at least 3). Otherwise the possibility exists the comparison system itself is under-performing itself or has a fault.

In order to check if the modules are performing according to the guaranteed specifications, an on-site test can be made with a so-called *peak watt meter* (these are, however, quite expensive, a question of hiring rather than buying). This will check the I-V curve of the modules and calculate the peak performance. If there is a problem with the modules, the manufacturer or supplier will need to be contacted.

1. Fault indication	2. Fault indication	Fault / possible cause / solution
System output less (per kWp) than a comparable system in same location	System not optimally designed	Inverter and array not very well matched. If not =>
		High losses / voltage drop in cables. Check calculations, possibly replace with larger cables. If not =>
	Incorrect installation	Strings not correctly wired, not plugged into connectors properly, loose connections, no voltage on terminals in PV array combiner box, incorrect DC polarity in circuit. Check for all of these. If not =>
	Modules not uniformly aligned, different tilts, orientations	Mismatch losses due to non-optimal design or installation. Visual inspection. Install another inverter (multi-string) or several inverters
	Array shading	Remove cause of shade if possible. If not, install multi-string inverter / several inverters. If not =>
	No clear indications	Inverter overheating due to clogged vents, bad ventilation and is derating itself. Clean inverter. Relocate inverter? Improve room ventilation? If not =>
		Check for possible problems originating on the grid. (see 5.4.2 *Problems originating on the utility grid*). Contact inverter manufacturer or utility. If not =>
		Modules have a lower peak performance than guaranteed by the manufacturer. Test with peak watt meter required. Module replacement? Compensation?

5.5.3 Malfunction category 3: System output lower than it used to be

The level of solar radiation can vary considerably from day to day, from week to week and from one year to the next – by as much as 20% (±10% from long-term annual average) to the year immediately preceding. So it is worthwhile comparing the output of the system being investigated with similar systems in the same locality (see previous). If variations in solar radiation can be ruled out as the cause of low output, refer to the troubleshooting table for this category.

1. Fault indication	2. Fault indication	Fault / possible cause / solution
The output of the system is lower than it used to be	The array current is lower than would be expected under high conditions of solar radiation, peak current is lower than previously. Power output is lower than it was	Check if the array is shaded or if there is dirt on it. Remove source of shade or clean. If not =>
		Defects in modules. strings, cables caused by storms or lightning etc.? Visual inspection. Check strings in PV array combiner box – Voc, Isc, I$_{MPP}$. Take measurements in conditions of constant sun, not in changeable conditions. Ideally also test with peak watt meter and compare with measurements made during commissioning. If not =>
		Disconnected terminal? Loose connectors? If not =>
		Defective bypass diodes in individual modules – caused by lightning / voltage surge? Short circuited diodes bridging over cell strings and reducing module output. Use process of elimination – first strings, then modules. If not =>
		Damage to module or cells caused by lightning. Cell damage may not be visible. Take module output readings. Replace module. If not =>
		Short circuit in module junction box due to moisture ingress. Rectify. If not =>
		Module degradation. Take readings from strings and modules and compare with data sheet. If not =>
	No faults or unusual readings detected in the array	Inverter overheating due to clogged vents, bad ventilation. Clean inverter. Relocate inverter? Improve room ventilation? If not =>
		Check for possible problems originating on the grid (see Section 5.3.2 *Problems originating on the utility grid*). (Data loggers useful in situations like this, check data and any faults registered). Contact inverter manufacturer or utility

6 Stand-alone PV Systems

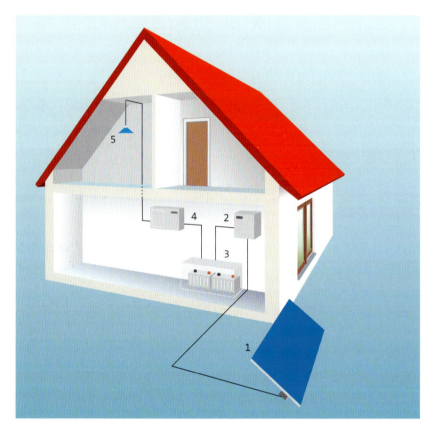

Figure 6.1: 1 PV array, 2 Charge controller, 3 Batteries, 4 Inverter, 5 Loads. Though a common configuration, many others are possible

6.1 Introduction

This chapter is an introduction to the design and installation of stand-alone PV systems. These are autonomous PV systems not connected to the grid. They range in size from pocket calculators to tens of kilowatts. The range of possible configurations is also considerable. Some stand-alone PV systems, known as *PV-hybrid systems,* also have another power source (a diesel-, gas- or biofuel generator or a wind or hydro-generator). Applications are varied – from airfield landing lights to telecommunication systems, not to mention satellites; and so are geographical locations – from the Sahara to the Arctic. The type of stand-alone PV system discussed here are mainly systems from about 40 Wp to about 1 kWp. The chapter will not deal with the design and manufacture of small integrated systems such as portable solar lanterns nor, at the other end of the scale, will it deal with the design and installation of very large systems (> 1 kWp) or hybrid systems. Readers wishing to design and install stand-alone PV systems will need more information than can be presented here – see 7.6 *Sources of further information* for suggested additional reading. National electrical codes also need to be referred to and installation manuals consulted The market for stand-alone systems and marketing them is discussed in 1.5.6 *The market for stand-alone photovoltaics*. The cost structure of stand-alone systems is different from that of grid-tied systems. The PV modules are typically between 30 % to 50 % of the cost of a system. Batteries usually have to be included in the cost estimate, but possibly also low energy and DC appliances. The fact that most systems are in remote locations also adds to the cost.

6.1.1 Stand-alone PV – basic description

Stand-alone PV systems are also known as *autonomous systems* or *island systems*. In a typical stand-alone PV system the DC electricity produced by the module(s) is used to charge batteries via a solar charge controller. If DC appliances are to be powered by the system, they are usually connected to the battery via appropriate-sized fusing, although some charge controls also provide limited appliance current. DC lights are usually connected to the charge controller. If AC mains voltage appliances are to be powered, this is done via an inverter connected directly to the batteries. The inverters used in stand-alone systems and the inverters used in grid-tied systems, though they both convert DC electricity to AC electricity, are very different devices and are rarely interchangeable. In some systems the DC electricity is used directly, without batteries. In pumping systems, water is pumped into a storage tank when the sun shines and used according to need. In stand-alone PV systems it is crucial that the energy produced is sufficient to cover the energy requirements of the loads. System voltages, i.e. the battery voltage or in the case of a system without batteries, the voltage of the loads, is usually 12 VDC, 24 VDC and less commonly 48 VDC.

In underdeveloped parts of the world a very common type of system is the so-called *Solar Home System* (SHS) consisting of one or two modules (40–100 Wp) powering lights and one or two small appliances. Where energy demand is higher, PV arrays consisting of several modules are needed and / or a PV-hybrid system with additional power sources.

6.1.2 Applications and limitations

The range of applications for stand-alone PV has already been discussed in 1.5.6 *The market for stand-alone photovoltaics* and the technology is proven. However, systems do need to be well designed and properly installed and it is important for designers and installers to be aware of the limitations of the technology.

Stand-alone PV systems of the type and range discussed here can be used very effectively in the following areas:
- homes – lights, radio, TV/video, computers, small refrigerators
- security – communications, lighting, alarms, CCTV
- medical – medical appliances, refrigerators, lights, communications
- office/work – lighting, computers, communications, ventilation
- water pumping – mainly for human consumption and livestock watering.

Although ideal for many applications, there are some inherent limitations to stand-alone PV:
- Systems need to produce the energy required by the loads at virtually the same time as it is needed. Energy can be and is stored in batteries but there are practical limitations to this – for example, to store energy produced in summer for use in winter would require a ridiculously large battery bank – even in a small solar home system.
- PV is not cheap – stand-alone PV usually makes more sense powering equipment (lights and all sorts of electronic equipment) which does not require large amounts of energy; powering heavy loads such as welders, X-ray machines, washing machines or shop equipment is possible theoretically but it is expensive and would require very large battery banks. With equipment in this power range, PV-diesel hybrid is an option. Appliances that produce heat are unsuitable because they simply consume too much energy, however there are alternatives.

Most stand-alone PV systems need to be managed – the exception being small systems with constant loads such as bus-shelter lights. Users need to know the limitations of a system and tailor energy consumption to basically how sunny it is and the state of charge (SOC) of the battery. However, a well-designed, well-installed and well-managed system should satisfy the user's electrical requirements practically all the time. Even the grid cannot guarantee that it will be on 100 % of the time.

If a grid supply is available (or not too far away), there is no economic advantage to replacing it with a stand-alone PV system. The actual cost of the electricity produced can be as high as € 2 per kWh. However, if the grid supply is unreliable, as it is in many developing countries, there can be a market for stand-alone PV even in places connected to the grid.

6.1.3 Future prospects for stand-alone PV systems

The future of stand-alone PV will also be affected by the foreseeable developments in PV technology discussed in 1.4.4 *Future potential*. The development of electricity storage technology (new and improved types of batteries and fuel cells) will be important. The increased availability of new low-energy lights and appliances – such as LED lighting should also have a significant impact, making stand-alone systems more economically attractive.

Figure 6.2: 50 Wp 12 VDC Solar Home System in Bangladesh. The system provides power for lights and a small water pump (Source: www.MicroEnergy-International.com)

6.2 Stand-alone PV system configurations

A range of stand-alone system configurations are possible, from the more straightforward to the relatively complex depending on power and energy requirements and electrical properties of the loads they are designed to power.

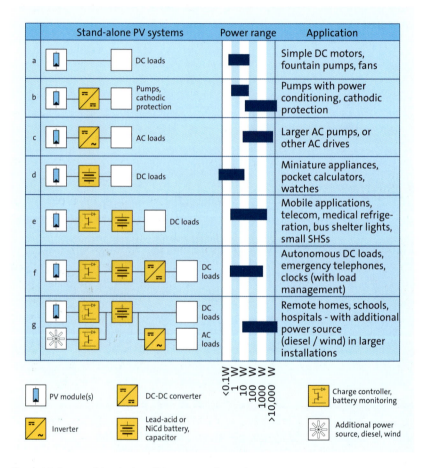

Figure 6.3: Range of stand-alone PV system configurations

6.3a shows a block diagram of a PV module coupled directly to a small DC motor, perhaps a solar fountain or small fan. The motor, as long as the module has been sized correctly, will run as long as there is sufficient solar radiation, its speed depending more or less directly on solar radiation intensity.

In larger systems, some sort of power conditioning is usually needed. In the case of a DC pump, it might include a current booster (for pump starting currents) and maximum power point tracking (6.3.b). Larger pumping installations may incorporate inverters to convert the DC electricity produced by the modules to that required by the AC pump (6.3.c).

Some form of energy storage is needed in most stand-alone PV systems. PV array output is not constant – the sun goes behind a cloud and output suddenly drops, sunny days are followed by overcast ones and there is no output at all at night. In miniature appliances, such as pocket calculators a capacitor is sufficient (6.3.d), but in anything larger a battery is needed, usually a lead acid battery of some kind, but because PV modules produce DC electricity and batteries store DC electricity and both are modular, systems can easily be assembled over a range a sizes.

The system configuration shown in 6.3e is typical for a so-called Solar Home System (SHS). Besides the charge controller, whose main function is to protect the battery, there is a DC-DC converter to power DC appliances which run on a different voltage to the system voltage (usually 12 or 24 VDC) (6.3.f), but inverters which will power AC appliances are much more common (6.3.g). Both of these will involve some power losses and it is worth considering the use of 12 or 24 VDC appliances. A considerable range is available. These are the types of systems which will be discussed here.

PV-hybrid systems are systems which have another power source besides the PV array (also 6.3.g). This could be wind, hydro or a generator (diesel, gas, bio-diesel). These systems can make particular sense when there is a large seasonal variation in levels of solar radiation. Some systems are powered mainly by PV in summer and mainly by wind in winter. When larger loads and appliances have to be powered, PV-hybrid systems with diesel generator back-up makes sense.

6.3 PV stand-alone systems – principal components

6.3.1 PV modules and arrays in stand-alone systems
The PV module(s) in a stand-alone PV system must produce enough electrical energy to power all the electrical appliances in the system and, in addition, enough to cover inefficiencies in batteries, inverters and cables. Unlike grid-tied systems, PV modules in stand-alone systems usually do not have MPP tracking, except in larger systems. This means that they generally have a much lower overall system efficiency.

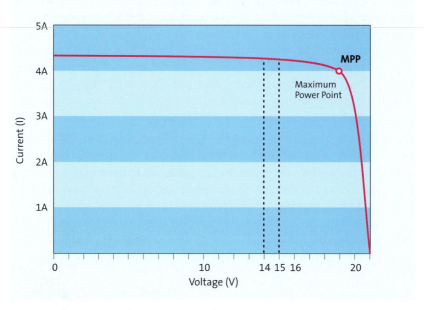

Figure 6.4: The operating voltage of a PV module in a stand alone system is typically 14-15V. This means that the module does not operate at the maximum power point (MPP) of the module I-V curve. MPP tracking can be used in stand-alone systems, but it is usually found only in large systems and pumping applications.

The PV modules need to be configured to match the system DC voltage, which is determined by the battery. System voltages are usually 12 VDC or 24 VDC and, on larger systems, 48 VDC. The operating voltage of the PV module(s) in a stand-alone PV system must be high enough to charge the batteries. A 12 V battery needs a voltage of 14.4 V to charge it. The PV module(s) must deliver this voltage to the battery after power losses/ voltage drop incurred in cables and across charge controller and diodes and often under conditions in which the solar cells are operating at high temperatures, up to 70 °C at times (see 2.3.5 *Effect of temperature on solar cell output*). A module with a V_{oc} of about 20 V is required to reliably charge a 12 V battery.

Modules chosen for charging batteries in stand-alone systems need to have sufficient cells in series to deliver the required voltage. For a 12V battery this is 36 crystalline silicon solar cells, and for a 24 V battery this is 72 cells (or 2 × 36 cell module connected in series). Or put another way, 6 cells connected in series for every 2 V battery cell.

Before the advent of the large grid-tied PV market, most PV modules were used to charge batteries and a nominal voltage (12 VDC, 24 VDC) was often given on the data sheet or the module label. But these days it is advisable to check that the modules are suitable for battery charging. Modules with junction boxes rather than those with plugged leads (which are used in grid-tied systems) are usually more convenient as the modules in arrays for stand-alone systems are connected up in parallel or series-parallel rather than in long series strings.

Most of the information regarding PV cells and modules in Chapter 2 *PV cells, modules and the solar resource* is also relevant for stand-alone systems. In most cases, because of the steeper angles required for best wintertime charging, and the fact that installations tend to be in rural areas rather than urban and there thus being more space available, PV arrays for stand alone-systems tend to be ground mounted on free-standing arrays, or, in the case of small numbers of modules, pole mounted. See 3.4 *PV module mounting structures and systems* for a discussion of mounting structures, in particular Free-standing mounting structures. Arrays need to be shade-free and well ventilated.

Figure 6.5: PV arrays for stand-alone systems are generally smaller than those for grid-tied systems and more likely to be mounted on free-standing structures. This one, at the KARADEA Solar Training Facility in Tanzania, is a manually operated tracking mount (Photo: Frank Jackson)

6.3.2 Charge controllers

The function of the charge controller (also called *charge regulator* or simply *regulator)* is to protect the battery and ensure that it has as long a working life as possible without impinging unnecessarily on system efficiency. Batteries are very sensitive to being over-discharged (see *Battery lives – cycle life and temperature* in 6.3.3 *Batteries)* and over-charging. Charge controller manuals should always be carefully read. Inappropriate use and incorrect installation can damage both batteries and controllers.

Figure 6.6: Charge controller (Source: Steca, www.steca.de)

The main functions of the charge controller are to:
- protect the battery from over-discharge – usually a *low voltage discon-nect* (LVD) disconnects the battery from the loads when battery voltage reaches a level which indicates that it has reached a certain depth of discharge (DOD)
- protect the battery from over-charging by limiting the charging voltage – this is particularly important with sealed batteries – here it is called *high voltage disconnect* (HVD)
- prevent current flowing into PV array at night. so-called *reverse current*.

Charge controllers also have other important functions, but not all charge controllers will have all of them; it depends mainly on size and price. Charge controller data sheets will give details. Among these functions are:
- different (manually adjustable) charging regimes for different battery types – sealed/gel batteries can be damaged if the charging voltage is too high
- overload and short circuit protection
- temperature compensation – tailors battery charging voltage to battery temperature, sometimes with external battery temperature sensor
- integrated lightning protection
- indicators – battery storage of charge, PV charging current etc.
- ability to deliver a regular equalization charge to flooded/wet cells/batteries (see 6.3.3 *Batteries*).

Some charge controllers will incorporate MPP tracking which will ensure that modules are operating optimally. These can increase output by 10% and more, but they do add significantly to the cost.

6.3.3 Batteries
The generation of electricity in stand-alone PV systems is rarely in phase with the power requirements of the loads. Lights and appliances need to be powered not only when there is sufficient solar radiation but also at

night and during periods of overcast weather, which can sometimes last for several days. The rechargeable batteries in stand-alone PV systems need to have long working lives under conditions of daily charging and discharging. Batteries which can do this are known as *deep-cycle* batteries. They also need to have a good charging efficiency at low charging currents and a low self-discharge rate. Choosing the appropriate type and correct size of battery is essential to ensure that a stand-alone PV system performs in the way it is required to perform and will also determine the working life of the battery.

A note about terminology: the term *battery* actually refers to a row of electro-chemical cells connected in series. These are sometimes simply called *cells* or *battery cells*. Several batteries together are usually referred to as a *battery bank*. An *accumulator* is another word for a rechargeable battery cell, but it is not used much these days.

The type of battery used in stand-alone PV systems, mainly for reasons of cost, is almost invariably the lead-acid battery and the comments made here only apply to this type of battery. A lead-acid electrochemical cell has a nominal voltage of 2 VDC. Six of these cells connected in series gives us a 12 V battery. Lead-acid batteries are available as 2 V cells, or as 6 V or 12 V batteries. By connecting these in series or parallel or series-parallel, required systems voltages – 12 V, 24 V or 48V – can be obtained.

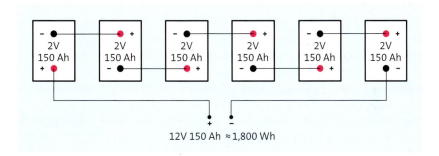

Figure 6.7: 1 Cells connected in series, 6 × 2 VDC to give a 12 VDC system voltage

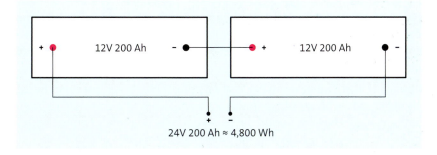

Figure 6.8: 2 Batteries in series, 2 × 12 VDC to give a 24 VDC voltage

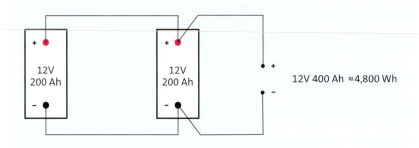

Figure 6.9: 3 Batteries in parallel, the voltage remains the same, 12 VDC

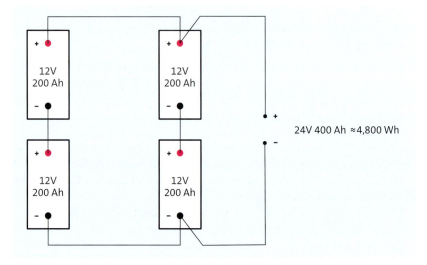

Figure 6.10: 4 Batteries in series-parallel to give 24 VDC

Stand-alone PV system voltage is determined by the choice of battery and the way the cells / batteries are configured. Cells should never be connected up in parallel and one should also avoid doing so with batteries. However, sometimes it is unavoidable if the right size and type of battery are not available, as is sometimes the case in remote areas or in the developing world. Series configurations are always preferable to parallel connection. In a series-connected row of batteries a faulty battery will be more easily recognized by a drop in overall battery bank voltage. In a parallel configuration this will not occur and the faulty battery will drain the good ones.

Physically, the lead-acid 2 V cell consists of positive electrodes and negative electrodes (plates of lead-alloy) in a container of electrolyte (diluted sulfuric acid). These are available as actual cells or already connected together in a single container as batteries, usually in units of 6 or 12 V.

Figure 6.11: Range of lead-acid cells/batteries of different sizes and suitable for stand-alone PV systems (Source: www.exide.de)

All makes and models of rechargeable batteries have been designed for specific applications and their electrical characteristics are different, but it is possible to broadly categorize them according to construction and the application they are designed for:
- SLI (start, lighting, ignition) batteries i.e. car and truck batteries
- stationary batteries – used in telecom, back-up etc.
- traction batteries – used in electric vehicles.

These 3 categories can also be sub-divided into:
- *flooded cells*, also called *wet cells*, and so called because they have a liquid electrolyte which needs to be regularly topped up with de-ionized/distilled water
- *sealed* batteries or *maintenance-free* batteries – which usually have either a gel electrolyte (hence *gel cells)* or incorporate a method of recombining the hydrogen and oxygen and do not need to be topped up, hence the term *maintenance-free*.

Batteries are also categorized by the type of plates they have – their physical construction and the composition of the lead-alloy. Flooded lead-acid batteries with tubular positive plates have the longest lives in stand-alone PV systems.

Automotive batteries (SLI) are designed to deliver heavy starting currents for a short period of time, not for the cycle regime of a stand-alone PV system or deep cycle use, so are not suitable. However larger truck SLI batteries are used with some success in small systems (< 100 Wp) where no other alternatives are available, such as in rural Africa. Modified SLI batteries are also often used in small systems in developing countries. In portable solar lanterns nickel-cadmium batteries are sometimes used.

Usual type description	Modified SLI	Gel cells, maintenance-free	Maintenance-free deep cycle	Flooded deep cycle
Construction	Thicker plates than SLI (automotive)	Maintenance-free, sealed	Gel electrolyte, tubular plates	Liquid electrolyte, tubular plates, transparent containers
Properties	Moderate to low water loss, low self-discharge rate	No maintenance	Low maintenance, can withstand deep discharge	Low maintenance, robust construction, charge well with low currents, can withstand deep discharge
Unit voltages	12 V	12 V	2 V–6 V	2 V–6 V
Capacity range in Ah	60–260 Ah	10–130 Ah	200–12,000 Ah	20–2,000 Ah
Self-discharge rate – monthly	2–4 %	3–4 %	< 3 %	2–4 %
% DOD – cycle life (approximate)	20 %–1000 40 %–500	30 %–800 50 %–300 (can be less)	30 %–3000 80 % > 1000	30 %–4500 80 % > 1200
Maintenance periods	3 months approx.	None	Monitoring & yearly cleaning	3 month approx.

An overview of lead-acid battery types. The important characteristics are cycle life and maintenance requirements. There are many types and sizes of batteries available and every model/make of battery will have different characteristics. Terminology also differs. The choice of battery is crucial and it is advisable to source them from PV system suppliers or suppliers who have experience of PV

Battery voltage during charge/discharge and measuring the state of charge (SOC)

An electro-chemical cell or battery is charged by applying a voltage higher than its nominal voltage to its terminals. The voltage of a lead-acid cell varies from its nominal 2 VDC depending on the SOC and on whether it is being charged or discharged. Voltage drops when a battery is being discharged, the drop being relative to the discharge current. When a battery is being charged, the voltage rises. When it is at rest, i.e. it is not being charged or discharged, the cell voltage will be somewhere between 2.00 and 2.12 V. So, in order to ascertain the state of charge (SOC), a voltage measurement can be taken. However, the battery needs to have been at rest for at least 4 hours beforehand. The voltage of a fully charged cell should be 2.12 V at 20 °C, for a 50 % charged cell 2.03 V and a cell which is completely discharged 1.96 V. For a 12 V lead-acid battery, these figures are simply multiplied by 6, the number of cells in the battery. A voltmeter can be used to estimate battery SOC, though it is not as exact as using a hydrometer (see *Using hydrometers* below). A more expensive meter that actually tracks amp-hours in and out of the battery for daily state of charge monitoring is recommended on any large system.

Equalization charges

During charging, at about 2.4 V (cell voltage), hydrogen and oxygen bubbles begin to form on the plates of lead acid-cells, the battery begins to gas and the water content of the electrolyte is reduced. Hydrogen gas is highly flammable. In order to avoid this, the charging voltage should not usually exceed 2.3 to 2.4 V per cell. In a stand-alone PV system the charge controller prevents this happening as it can lead to damage. However, occasional overcharging of flooded cells is beneficial as it mixes up the electrolyte and reduces its stratification into different levels of density/acid concentration. This is called an *equalization charge*. Good quality charge controllers and inverter-chargers will do this automatically. Sealed/gel batteries should never be overcharged (see HVD in 6.3.2 *Charge controllers.*)

Best practice battery charging

Ideally, batteries are charged in 3 stages. During the *bulk charge*, the battery is given as large a charge as possible. Then it is given an *absorption charge* at a constant voltage but at reduced current to complete the charging process. The last stage is the *float charge*, in which the battery is basically just kept topped up. Good charge controllers will more or less charge batteries like this, but are dependent upon solar radiation levels. Sometimes it is a good idea to occasionally disconnect all loads from a system to allow it to have a good charge. In larger PV-diesel hybrid systems, the inverter-charger will usually ensure that this is done correctly. Properly charging the battery will help it to retain its capacity and last longer.

Using hydrometers

The SOC of a flooded cell can be measured using a hydrometer to measure the specific density of the electrolyte – the higher the SOC, the higher the concentration of acid. During discharge the sulfur ions bond to the lead plates leaving more and more diluted acid behind, so that what is left becomes closer to plain water. Battery data sheets should have a table giving exact values and the measurements taken can be compared to these. The battery also needs to have been disconnected from power sources and loads for several hours beforehand. Typical values for a lead-acid battery are 1.28 g/cm³ when fully charged, 1.18 g/cm³ at 50 % SOC and 1.10 g/cm³ when fully discharged. The acid used in deep-cycle batteries is weaker than the acid used in common SLI batteries. So if the hydrometer has red, yellow, and green zones on the float, ignore them. A deep-cycle battery will hardly ever get into the green and that's just fine. A useful rule of thumb for estimating cell voltage: cell voltage = acid concentration + 0.84. However, a hydrometer cannot be used with sealed batteries, nor is it recommended for regular use on wet-cell batteries. The fewer times battery caps are removed, the safer one's clothes and the battery will be. Use the hydrometer for a once or twice yearly state of health monitoring.

Capacity of lead acid batteries

The electrical storage capacity of a battery is measured in amp-hours (Ah), or ampere-hours. This describes for how many hours a specific current can be delivered by a fully charged battery before it is discharged. This can roughly be converted into watt-hours (Wh) by multiplying the battery Ah by its voltage, but only very roughly. This is because, at low discharge currents, significantly more electricity can be delivered by a battery than at high discharge currents. For example, a battery that will deliver 1 A for 100 hours has a capacity of 100 Ah. However, if the same battery is delivering a current of 8 A, it may do so for only 10 hours, giving it a capacity of only 80 Ah. This is the reason that capacity is always given as Ah at so many hours. For instance, 100 Ah at 100 hours or 80 Ah at 10 hours. Capacities at 20 or 100 hours are the usual standards in the solar industry, but one may find other listings. Just be sure to compare like to like.

The capacity of a battery is affected by the current at which it is discharged and its temperature. At low temperatures the capacity is significantly reduced. As a rule of thumb capacity is reduced by 1 % with each °C drop in temperature. If the manufacturer does not give a temperature for battery capacity, it can be assumed to be for an operating temperature of 20 °C. Example: a fully charged (100 %) battery at 20 °C will only have 80 % of its capacity at 0 °C.

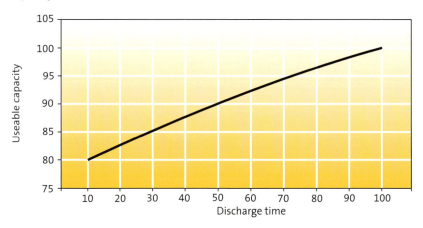

Figure 6.12: Example of relationship between discharge time and capacity of a lead- acid flat plate battery

Nominal capacities for batteries are given for particular rates of discharge. This is expressed in the form of C-rates. The C-rate refers to the number of hours a battery will deliver a specific current. C_{100} refers to a period of 100 hours, C_{20} refers to 20 hours. For example, a 100 Ah battery at C_{100} will deliver 1 A for 100 hours. However, if the same battery is discharged at a higher current over a shorter period of time, say 8 A for 10 hours, it will have a lower capacity, 80 Ah at C_{10}. On the battery data sheet two capacities would be given: 100 Ah battery at C_{100} and 80 Ah at C_{10}. A lot of batteries are rated at C_{10}, but, for most PV applications, the C_{100} rate is more important. The data sheets for quality batteries will give capacities for a range of C-rates.

Battery lives – cycle life and temperature

The chemical reactions in batteries are not fully reversible. Every time a battery is discharged small quantities of lead-sulfate are created. This is called *sulfation*. After every discharge the battery degrades slightly and it loses capacity. The more often it is discharged and the deeper it is discharged, the greater the degradation. Charging and discharging a battery is known as *cycling* the battery, and the number of cycles a battery can go though – usually one a day, in stand-alone PV systems – is known as the battery's *cycle-life*. This needs to be taken into consideration when sizing batteries, as it will determine how long the batteries will last.

Batteries should never be fully discharged (see LVD in 6.3.2 *Charge controllers*). They should never be brought below their *maximum allowable* depth of discharge DOD. For a deep-cycle battery, this is typically 80 % DOD. The *recommended* DOD is less, and this is the DOD recommended for the battery to have a reasonably long life, usually about 50 % for a deep-cycle-flooded battery. If it were regularly discharged to only 20 %, it would last longer still. Lead-acid batteries will slowly discharge themselves if left standing. The self-discharge rate of batteries typically used in PV systems is about 3 % per month.

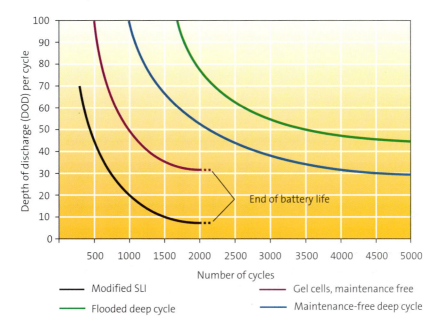

Figure 6.13: The cycle lives of four different batteries in relation to depth of discharge (DOD), Each make/model will have its own cycle life curve. Shallow cycles make longer life expectancies regardless of battery type

Temperatures under 0 °C can destroy lead-acid batteries as the electrolyte can freeze. As batteries discharge, the acid concentration drops and the electrolyte will freeze at a higher temperature. A fully charged battery can withstand temperatures of -40 °C. A 50 % discharged battery can freeze at -10 °C. Too high a temperature can also cause damage. The life of a battery which is regularly operated at 30 °C can be only half that of one operated at 20 °C. Temperature should not be allowed to exceed 55 °C as that will cause permanent damage.

Battery Ah efficiency

The Ah efficiency of a battery is the relationship between the number of Ah put into the battery and the number of Ah that can be taken out, which is always less. In a new deep-cycle battery this will be 90 % under ideal conditions of charging regime and temperature, but in real systems a return of 75 % of the energy put into a battery is good.

Choosing the appropriate battery

There are a large number of batteries available on the market, each one designed for a specific application. Choosing a battery for use in a stand-alone PV systems means assessing the relevant characteristics and making a decision on that basis. The important characteristics are:
- price-performance relationship
- capacity
- cycle life
- maintenance requirements, if any
- size and space required
- Ah efficiency
- self-discharge rate
- for mobile applications (vibration proof? sealed? can it be installed non-vertically?)
- environmental aspects (suitability for use near water supplies or in nature reserves?).

In smaller systems, the choice of battery will usually be 12 V flooded or gel cell units followed by 6 V units. Sealed batteries are preferable in mobile applications such as recreational vehicles. In larger systems, those with PV arrays above 1kWp or incorporating an inverter or inverter-charger above 1 kW, 2 cell units with tubular plates are advisable, flooded-cell deep-cycle units being usually the first choice. Sealed/gel batteries have the advantage of not requiring regular maintenance and are usually not considered to be hazardous loads, however this does need to be verified with transport companies. The choice of battery is crucial for the success of a stand-alone PV systems. System size, application, location and code requirements all need to be taken into consideration.

6.3.4 Electrical appliances, lights and loads

Electricity generated by PV is expensive so it is especially important to use low energy and energy efficient appliances. These may cost more than less energy efficient equivalents, but overall system costs will be considerably lower. The power or watts (W) of appliances is important but what is more crucial for PV array and battery size is the energy or the watt-hours (Wh) they consume – the power (W) multiplied by the time in hours (h) that an appliance is running (see *Step 2 – System energy requirements* in 6.5.4 *Sizing example – holiday home* for worked examples). The power consumption figures on the labels of appliances are usually accurate enough. The energy required each day to run the appliances will not only determine the number of PV modules and batteries needed, but will also affect the life of the batteries. Some systems power only DC appliances (usually small lighting systems), others will have both DC and AC loads and larger systems tend to power mainly AC loads.

Lighting is a major application in stand-alone PV. Compact fluorescents (available in both DC and AC versions) are recommended. They have a high *efficacy* (light output versus watts). LEDs also have very high efficacies combined with exceptionally long life expectancies. Halogen lamps are better than standard incandescents, but still have low efficacies in comparison to compact fluorescents and are generally not recommended. Incandescent lamps have very low efficacies and should be avoided.

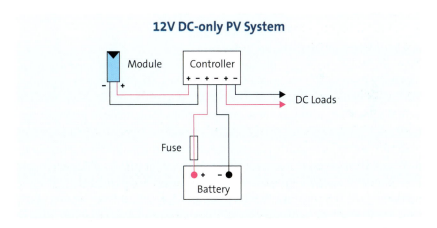

12V DC-only PV System

Figure 6.14: Wiring diagram of a simple 12 VDC stand-alone PV system. The loads, usually lights in systems like this, are connected directly to the charge controller. Note location of main battery fuse. For an example of a large DC/AC system with inverter see 6.5.4 Sizing example – holiday home

DC TV-videos, sound equipment and other appliances are available. They are used in recreational vehicles, caravans and boats, more often at 12 VDC than 24 VDC. Since they are designed to be run by batteries, they are usually quite efficient. Two-way radio is nearly always run on DC.

Desk top computers (with cathode ray tube monitors) are high on energy consumption. Flat screens are better and laptops have the lowest power requirement – 20 W is not untypical. DC-DC converters are available for running cell phones and laptops directly from batteries. This is more efficient than using an inverter to convert to AC, then plugging in a standard charger to convert back to DC again. Extra conversion steps always cost energy. Avoid them when possible.

Very high efficiency 12 / 24 VDC refrigerators are also available. These and other high-surge appliances are usually connected directly to the battery, via appropriate fusing of course. However, some caution is called for – high starting currents can interfere with the working of some charge controllers – the battery voltage will momentarily drop and the charge controller might interpret this as a low battery. This could be a problem if lights are connected directly to the charge controller and the charge controller switches them off to save power. The issue may need to be discussed with the charge controller supplier or manufacturer.

Systems which are DC-only have their limitations. The difficulty and expense of getting DC appliances (and sometimes the problem of having to use long cable runs and thus cables of a large cross-sectional diameter because of voltage drop – see 7.7 *Power losses and voltage drop in PV systems*) means that using an inverter is usually cheaper and more convenient. A good inverter will have no problem running most low to medium power home and office appliances in a stand-alone PV system.

6.3.5 Inverters in stand-alone systems

PV modules and batteries produce DC electricity. However, most electrical appliances run on AC. DC appliances are available but most system owners also want to use normal AC home and office appliances. Inverters used in stand-alone PV systems are very different from those used in grid-tied systems and are sometimes called *battery-based inverters* or *island inverters*. Inverters are rated in watts (sometimes in volt-amps) and the nominal rating should be sufficient to power all the AC appliances that will be on at any one time. When powering motors and other loads which require high start-up currents, it is important that the inverter can deliver the required start-up current. Inverters in stand-alone PV systems also need to be efficient.

Inverters in stand-alone systems should have the following characteristics:
• produce a pure AC sine wave (preferably)
• stable AC voltage and frequency output
• have a DC input range that takes into account changing battery voltages
• be sized to provide sufficient continuous power for all the loads
• surge capacity to deal with high starting currents if required
• high efficiency at full and part load
• high level of reliability
• low electromagnetic interference
• low power consumption in stand-by mode.

Inverters do shut down if battery voltage gets too low, but this is to protect the inverter, not the battery. This means that stand-alone PV systems need to be monitored manually or with a remote monitor/Ah meter to ensure that batteries are not over-discharged/and inverters are rarely *reverse DC polarity protected*. This means that they can be destroyed if connected up incorrectly to the battery (see 6.6.2 *Electrical work on stand-alone PV systems and particular hazards*).

More than one inverter can be used in a system, however each one needs to have its own dedicated output distribution circuit(s) – exposing the AC output terminals of one inverter to the AC output voltage of another inverter will usually result in damaging or destroying both of them.

However, some more sophisticated inverters are designed to be run in parallel and synchronized with each other in a *master-slave* setup. This naturally gives a higher level of power security; and on larger systems can increase efficiency by avoiding running both inverters on part loads. The *master* inverter usually supplies all the loads; while the *slave* comes on line only when required.

Inverters incorporating other system components

Recently several devices have come on the market which combine an inverter with other system components, such as a solar charger controller, a battery charger and load distribution control unit. These can reduce installation costs and unauthorized interference with the system; however, because all the components are pre-sized, the designer cannot exercise much discretion in selecting and sizing components as would normally be the case.

Inverter-chargers combine an inverter and a battery charger. They can be used when there is a another AC power source (diesel generator or grid) in addition to the PV array, such as on boats (shore power) and in emergency power supplies and PV-diesel hybrid systems. These devices are quite sophisticated and in addition to operating as an inverter and battery charger, can operate as uninterruptible power supplies (UPS) in parallel with each other and the grid. While inverter-chargers can be connected up to the grid, this does not mean they are *grid-tied inverters* – they only take power from the grid, they do not feed power onto it.

6.3.6 Cables and accessories

Array cables need to be UV resistant and suitable for outside application. Much that has been said in 3.3.1 *Cables,* 4.4 *DC cable installation – general guidelines* and 4.4.1 *DC wiring – modules and strings* is relevant, however voltages on stand-alone arrays rarely exceed 48 VDC. It is essential to keep power losses/voltage drop in cables to a minimum – less than 3 % between the array and the batteries and less than 5 % between battery and DC loads (codes need to be referred to). Inverters are usually supplied with battery-inverter cables or the manual will specify cable sizes – they are usually quite thick, a 2 kW inverter with 24 VDC input will draw a current of over 80 A from the batteries. National codes will specify the type of cables to be used in AC and DC distribution circuits. See also 7.7 *Power losses and voltage drop in PV systems*. Cables in battery rooms may have additional specifications, i.e. resistance to acid.

If the diameter of the cables from the PV modules to the batteries is not sufficient, the life of the batteries can be adversely affected, particularly if the operating temperature of the solar cells is high (as high as 70 °C in some conditions). Permissible voltage drop will be exceeded and the batteries will not be fully changed.

On systems having more than two modules, a DC disconnect/isolator should be installed between the PV array and the charge controller so that the system can be isolated from the array while any work is being done on it or during lightning storms

Main battery fuse

The main item of electrical safety equipment in a stand-alone PV system is the main battery fuse to protect against battery short circuit. This needs to be DC rated and have an adequate breaking capacity. This should be installed as near as possible to the positive pole of the battery and should be sized to protect all components that could be affected by a short circuit from overload. Suppliers of quality inverters usually offer a range of appropriate fuses.

A battery short circuit is extremely dangerous. The extremely high current that can flow can cause sparks and arcing, cables can melt, the battery can explode splattering battery acid everywhere. Battery acid can cause blindness and burns and inhalation of acid vapors can cause lung damage. Put an appropriate-size fuse or circuit breaker on ANYTHING connecting to a battery!

6.4 Designing stand-alone PV systems

A systematic approach is important when designing and sizing stand-alone PV systems. The procedure is more or less as follows:

- feasibility assessment, site survey, including initial assessment of solar resource at site
- assessing energy requirements, including potential energy saving options
- system concept development
- more detailed assessment of solar resource at site
- sizing the main system components
- selecting the main system components
- review of system design and size.

6.4.1 Feasibility assessment, site survey

The first thing to do is to assess the feasibility of the project the customer has in mind. It may be straightforward, compromise solutions might need to be found or a stand-alone PV system may not be the solution. The solar resource at the site may be known to the contractor if it is in their usual area of operation, but, if not, it may need to be researched at this stage (see 6.4.4 *Assessment of solar resource for stand-atone PV system site*). The feasibility assessment/site survey is a very important first stage of any project and has been discussed extensively in 1.5.3 *Initial consultancies with customers*, 1.5.4 *Contracts, quotes, estimates and insurance*, 1.5.6 *The market for stand-alone photovoltaics* and 1.5.7 *Opportunities for marketing other renewable energy technologies*. If the project is going ahead, the next stage is to collect all the information required. Ideally a visit should be made to the site, but sometimes the site may be a long distance away – even on another continent – stand-alone systems are usually installed in remote locations. If this is not possible, many installers get potential customers to fill out a form and provide photographs of the site – see 7.5 *Site survey form – stand-alone PV systems* for an example. What needs to be done at this stage is to:

- clarify exactly what the customers wants the system to do
- clarify if any later expansion of the system is planned
- make an assessment of system energy requirements
- establish the pattern of system use – all year round, seasonal, weekends?
- establish the degree of power security needed
- assess installation logistics.

6.4.2 Assessment of system energy requirements

One of the first things to do when designing or assessing the feasibility of a stand-alone PV system is to estimate energy requirements. This is usually done in terms of watt-hours required per day (Wh/day) – doing this in days fits in neatly with the solar cycle and is useful for later battery sizing. How and when the system is to be used is also very important. If it is only going to be used at weekends for example, the energy requirements can be spread out over the week. It is also useful to look at non-electrical options for appliances that use large amounts of electrical energy. An electric kettle for example, could be replaced with a gas kettle. In the case of water heating, PV will prove to be too expensive in any case, but there is always the option of solar water heating (see 1.5.7 *Opportunities for marketing other renewable energy technologies*). Replacing more conventional electrical lamps and appliances with low-energy versions is almost invariably more economic in terms of overall system cost. The use of DC appliances should also be considered.

A list of all the appliances to be used in the system needs to be made and their power rating (W) and the number of hours (h) they will be on each day noted. The energy rating of appliances can usually be obtained from the label or user instructions. If only the current (A) is indicated, multiplying by the appliance voltage (V) will give the wattage (W). If no details are available, the power rating can be assumed to be that of a similar type and size of appliance. Multiplying the power rating of the appliance by the number of hours it will be on will give the energy requirement: watts (W) × hours of use (h) = energy requirement (Wh) (see worked example in 6.5.4 *Sizing example – holiday home*). A glance at the list will show which appliances require most energy and alternatives can be considered. However, losses in batteries, inverters and cables also need to be taken into consideration – this can be done at this stage or later when assessing the amount of energy the PV array has to produce – in the sizing method used in this chapter, the latter approach is taken (see 6.5.1 *Sizing the PV array*).

6.4.3 System concept development
An overall design concept/system configuration needs to be decided on. Very often several options need to be examined before coming to a final decision. Doing rough drafts can be very useful at this stage. Also keep in mind that sometimes several small systems are preferable to one single large one. Small systems are easier for users to manage and maintain and if there is a problem, it does not shut down everything. This can be very important in remote areas where local expertise is limited.

Another thing to be decided at this stage is the system voltage: 12 VDC or 24 VDC or (in larger systems (48 VDC). Several items influence this decision:
- length of cable run from PV modules to batteries
- overall system size – smaller systems usually operate at 12 VDC
- if there is to be a large inverter in the system – inverters over 2000W are usually 24 VDC and very large inverters are often 48VDC
- long DC distribution circuits may require 24 VDC in order to avoid large cable diameters; high power consuming DC appliances will generate high currents, the current consumed by a 12 VDC 500 W appliance is 40 A
- availability of DC appliances (particularly lights) – some types of lights are only available in 12 VDC.

In DC-only systems the availability of the appliances at the appropriate DC voltage is essential. However, nowadays there is quite a range of DC appliances available from recreational home and caravan suppliers and there are some very good DC fridges and freezers on the market.

6.4.4 Assessment of solar resource for stand-alone PV system site
The optimum tilt angle for the PV array for the system design month at the site needs to be determined. Then a figure for solar radiation at that angle and during that month needs to be arrived at – in average daily kWh/m² for that month (peak sun hours). Methods of doing this have been discussed in 2.5 *Estimating the output of photovoltaic systems, array angles and orientation* and 2.6 *Sizing and design software*.

6.5 Stand-alone PV system sizing

6.5.1 Sizing stand-alone PV arrays

The following formula can be used to size the type of stand-alone PV system discussed in this chapter:

$$W_{PV} = E \div G \div \eta_{SYS}$$

where

W_{PV} = peak wattage of the array [Wp]

E = the daily energy requirement in watt-hours [Wh]

G = average daily number of peak sun hours in the design month for the inclination and orientation of the PV array

η_{SYS} = total system efficiency expressed as a factor (1 is 100 %) can be taken to be generally 0.6 (60 %) (see below for details of how this figure was arrived at).

The design month is the month with the highest ratio of load to solar radiation. It is usually the month of lowest average daily solar radiation during the period the system will be operational. It will be December or January if the system is going to be used all through the year and the energy requirement is fairly constant. However, in some circumstances it can be a month of high average daily solar radiation; for example, if a pump is required to pump a lot of water during the summer months or winter sun is insufficient and the customer agrees to use a generator occasionally to make up the winter energy shortfall. In northern latitudes, except for very small systems, designing for total PV support on winter sun makes for very large PV arrays that shut off early in the day in summer and is not a good use of capital.

Note that the number of peak hours is for the inclination and orientation of the PV array. If the only information available is for solar radiation on a horizontal plane, then a tilt and orientation correction factor will need to be applied to this figure (see also 2.5 *Estimating the output of photovoltaic systems, array angles and orientation* and 7.6 *Sources of further information.)* The easiest and most accurate way of obtaining a value for G is to use sizing software, but manual methods using maps can also be used and the degree of accuracy achieved can be acceptable (see worked example in 6.5.4 *Sizing example – holiday home).* Designers and installers should familiarize themselves with manual methods, if only to check the plausibility of results obtained using sizing software, at least roughly. And when using manual methods, it is not a bad idea to check the results of one using another.

A totally efficient system is not possible. There are losses in each component and components never match perfectly. An overall system performance ratio of 60 % is quite good. The efficiency factor of 0.6 used here has assumed quality components and correct cable sizing. It has been arrived at as follows:

$$\eta_{SYS} = \eta_{PV} \times \eta_{PV\text{-}BATT} \times \eta_{CC} \times \eta_{BATT} \times \eta_{DIST} \times \eta_{INV} =$$
$$0.8 \times 0.97 \times 0.98 \times 0.9 \times 0.97 \times 0.9 = 0.60$$

where

η_{SYS}	= total system efficiency
η_{PV}	= 20 %, to account for the PV modules not operating at MPP, hence 0.8 – see 6.3.1 *PV modules and arrays in stand-alone systems* for more details
$\eta_{PV\text{-}BATT}$	= 2 % losses due to voltage drop in cables from PV array to battery, hence 0.98
η_{CC}	= 2 % losses in a good quality charge controller, hence 0.98
η_{BATT}	= 10 % battery losses, battery Ah efficiency, hence 0.9
η_{DIST}	= 2 % losses in distribution cables from PV battery to loads, hence 0.98
η_{INV}	= 10 % losses in a good quality inverter, hence 0.9.

The quality of components such as charge controllers, inverters and batteries and the correct sizing of the DC cables in the system will affect overall efficiency. Most text books and good catalogues give slightly different sizing formulae. Some use amp-hours (Ah) instead of watt-hours (Wh). Results will be similar. It must also be said that it is very difficult to predict how a system will be used, so energy requirements are also difficult to predict exactly. In reality, users learn to manage systems and in systems where a loss of power is unacceptable, arrays and batteries are oversized and a back-up generator is mandatory.

6.5.2 Battery sizing

The battery needs to be sized to store not only the daily energy requirement, but also several days' extra. This is to provide energy during overcast days, to cover system energy losses and to ensure that the battery is not discharged beyond its maximum allowable DOD.

The following formula can be used:

Q	$= (E \times A) \div (V \times T \times \eta_{INV} \times \eta_{CABLE})$
Q	= *minimum* battery capacity required in amp-hours [Ah]
E	= the daily energy requirement in watt-hours [Wh]
A	= the number of days of storage required
V	= the system DC voltage [V]
T	= the maximum allowable DOD of the battery (0.3 to 0.9)
η_{INV}	= inverter efficiency – this is 1 if there is no inverter
η_{CABLE}	= the efficiency of the cables delivering the power from battery to loads

See 6.5.4 *Sizing example – holiday home* for a worked example of this formula.

The number of days of battery storage required also depends on how crucial it is that there is always enough energy available to power the loads. In reality it ranges from 3 days (small SHSs in Africa) to 20 days (found in some systems in North America). Three to five days is about average. Size for less than three days and the battery is going to be cycled heavily every day. More than five days gets seriously expensive. Starting the generator is usually a better choice. In a holiday home the users simply might have to be more careful during overcast periods, but in the case of a medical refrigerator used for vaccine storage, power needs to be guaranteed. Medical refrigeration systems are also usually autonomous and not connected to other loads which might drain the batteries.

The maximum allowable DOD of the battery depends on the type of battery. No lead-acid battery in a stand-alone PV system should ever be fully discharged. This figure is often given as a percentage. If the maximum allowable DOD for a battery is 80 %, then T = 0.8. It should be on the battery data sheet. See also *Battery lives – cycle life and temperature* in 6.3.3 *Batteries* regarding DOD and battery life.

 When battery cells are connected in series, the voltages are added together but the amp hours (Ah) remain the same. When they are connected in parallel, the voltage remains the same but the capacities (Ah) are added together. In both cases the amount of electrical energy (Wh) stored remains the same (see also diagram in 6.3.3 *Batteries*).

6.5.3 Cable selection
A particular issue with stand-alone PV systems is power loss and voltage drop in cables carrying 12 VDC and 24 VDC – see 7.7 *Power losses and voltage drop in PV systems*. See also 6.3.6 *Cables and accessories*.

6.5.4 Sizing example – holiday home

Step 1 – Site survey
A holiday cottage is to be used all the time from March to October (7 days a week). It will only be used intermittently in the winter months. It needs to provide power for lights, a TV, some kitchen appliances and a low energy AC refrigerator. It is located in Montana, USA, at close to latitude 47° N.

Step 2 – System energy requirements
The system owner wants to use only AC appliances for convenience reasons.

Loads/appliances	Power rating of appliances [W]	Quantity	Total power required [W]	Hours of use per day [h]	Daily energy requirement [Wh]
Low energy compact fluorescent lamps – living room and hall	11W	3	33 W	3 h	99 Wh
Fluorescent lamp – kitchen	20 W	1	20 W	1.5 h	30 Wh
Lighting – outside & shed	100 W	2	200 W	0.2 h	40 Wh
Color TV	60 W	1	60 W	1 h	60 Wh
Small microwave oven	700 W	1	700 W	0.4 h	140 Wh
Food mixer	400 W	1	400 W	0.1 h	40 Wh
Refrigerator (low-energy, compressor, 110 VAC)	80 W	1	80 W		400 Wh (from appliance documentation on refrigerator label)
Totals			1,493 W		809 Wh

Step 3 – System concept development

The nearest spot – big enough and shade-free – suitable to locate a free-standing ground mounted array is 30 m from the house. So, to reduce power losses, a system voltage of 24 VDC is chosen. An inverter will be needed to power the AC electrical appliances. A remote monitor/Ah meter will be installed in the main room of the house so that the owner can monitor the battery SOC and manage energy consumption.

Using the panel angle graph in 2.5 *Estimating the output of photovoltaic systems, array angles and orientation*, we can see that the best angle for this latitude for the crucial months of March and October is 50°. If the system were to be used all through the winter a steeper angle would be required.

Step 4 – Assessing the solar resource

The information we need is the number of peak sun hours at the angle we have chosen. Sizing software could do this, but in this case we will use information provided in a series of maps published by the NREL, National Renewable Energy Laboratory on their web site (http://www.nrel.gov/) and which have been condensed into a bar chart expressing the solar average daily insolation levels for each month – see 2.5 *Estimating the output of photovoltaic systems, array angles and orientation* for the bar chart. Consulting this bar chart, we note that the number of peak sun hours received at the angle of latitude (47° N) at this location is 4.5 in March and October. And in this case, since the tilt angle of the array will be 50°, this data is sufficient. But things are not always so ideal.

If the design month was in mid-winter, say December, the optimum angle would be about 70° in order to catch the low winter sun. The average number of peak sun hours per day in December from the bar chart is 2.5 *but* that is for a surface tilted at the angle of latitude (50°). However, a surface tilted at a steeper angle in that month, will get slightly more than 2.5, so the bar chart is still useful.

If the design month was in mid-summer, say May or August, the optimum angle would be about 35° in order to make the most of the high sun during the middle of the day. The average number of peak sun hours per day for May or August is 5.5, *for* a surface tilted at the angle of latitude (50°). On a surface at 35° it would in fact be more, so it would be easy to oversize the system.

These are some of the limitations of using maps. They have their uses with small systems, but results need to be checked and using sizing software makes the process easier and more accurate. Oversizing or undersizing a 100 Wp array by 10% is one thing, making the same error on a 2,000 Wp array is another.

Step 5 – Sizing the PV array and the batteries

Sizing the PV array
The formula from 6.5.1 *Sizing the stand-alone PV array* can be used:

$$W_{PV} = E \div G \div \eta_{SYS}$$

where

W_{PV} = peak wattage of the array required in Wp]

E = the daily energy requirement in watt-hours – 809 Wh

G = 4.5 peak sun hours

η_{SYS} = total system efficiency, taken to be 0.6.

W_{PV} = 809 Wh ÷ 4.5 peak sun hours ÷ 0.6 = 299.6 Wp

So the minimum size of the array will be 300 Wp. Later, when all components and appliances have been chosen, this figure is checked again – and confirmed by the system supplier.

Sizing the battery
The formula from 6.5.2 *Battery sizing* can be used:

$$Q = (E \times A) \div (V \times T \times \eta_{INV} \times \eta_{CABLE})$$

where

E = the daily energy requirement of the system, 809 Wh,

A = the number of days of storage required – 5 days' battery storage is judged to be adequate for the application and the location in this example.

V = the system voltage, in this case 24 VDC,

T   = the maximum allowable DOD for the battery chosen from the catalogue, 50 % so T = 0.5;

η_{INV} is inverter efficiency – in this case 90 %, so η_{INV} = 0.9,

η_{CABLE} is the efficiency of the cables in the distribution circuits – in this case, we are assuming it will be 3 % so η_{CABLE} = 0.97.

Q = (5 × 809 Wh) ÷ (24 V DC × 0.5 × 0.9 × 0.97) = 386 Ah

The minimum capacity of the battery needs to be 386 Ah. The system voltage is 24 VDC, so either 2 × 12 V batteries; or 4 × 6 V batteries or 12 × 2 V cells are needed.

Step 6 – Selecting the components

The catalogue of a PV system supplier is consulted. The location is remote and a module that is easy to handle is required. An 80 Wp crystalline module (8 kg, 1.2 m × 0.5 m) is chosen. It has a V_{OC} of about 21 VDC, i.e. a nominal voltage rating of 12 V. The PV array required is 300 Wp minimum, so 300 Wp ÷ 80 Wp = 3.75 modules are required, i.e. 4 modules. This gives an array size of 320 Wp – a little extra does no harm. Since the system voltage is 24 VDC, the array needs to be wired up in series-parallel. The modules have junction boxes on the back rather than plugged leads which facilitate easier series-parallel connection.

Now a charge controller needs to be selected. The current rating of the charge controller is determined by the current produced by the array and the current consumed by the loads. The current produced by the 80 Wp module is 4.5 A (from module data sheet). Pairs of modules are connected in series to get the 24 VDC configuration (12 V + 12 V) but the current remains the same – 4.5 A. However, we need 2 pairs to get 320 Wp, so the input current rating of the controller needs to be at least 9 A. No DC loads are envisaged. Maximum array current will be 9 A, but since system expansion is a real possibility, a 20 A charge controller is selected. To monitor the batteries, it is also decided to install a remote monitor/Ah meter that can be put on the wall in the main room.

The AC total power used in the system at any one time will be 1,493 W. A 2,000 W inverter with 24 VDC input is chosen. The extra 500W is to allow for system expansion and high starting currents for motors. It has a high efficiency at low loads, i.e. an efficiency of over 90 % when it is running loads of 10–20 % of its full load rating.

The battery closest to the minimum 386 Ah required in the catalogue is a 12 V battery with a capacity of 400 Ah. It is a maintenance-free deep-discharge gel battery with flat plates and the price is reasonable. Since the system voltage is 24 VDC, 2 of these batteries will be needed, connected in series to give the 24 VDC.

Final system design

So the system will consist of:

- 4 PV modules 80 Wp, nominal voltage 12 V
- 1 charge controller, 20 A input and output current
- 2 deep cycle 12 V 400 Ah gel cell batteries
- 1 inverter, 2000 W continuous power rating, 24V input.

24V/230VAC PV System with Inverter

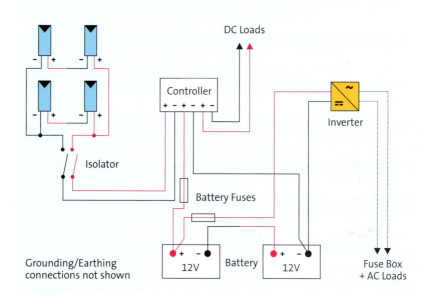

Figure 6.15: Wiring diagram of a DC/AC stand-alone PV system similar to the one designed. The AC voltage of the above system is 230 VAC. In North America it would be 120 VAC. Note that the inverter is connected directly to the battery and DC loads (optional) are connected to the charge controller. Note also the position of main battery fuses. The double-pole discon-nect/isolator between the PV modules and the charge controller enables the PV array to be disconnected from the system for maintenance, repairs and during thunder storms.

6.6 Installation and commissioning

6.6.1 General safety guidelines

When installing stand-alone PV systems, all the usual precautions taken when working on any electrical system need to be observed. The batteries in these systems present additional dangers and many electricians will not be used to working with DC electricity and the particular electrical characteristics of PV arrays. Further information on these issues is given below in 6.6.5 *Battery safety* and 6.6.7 *Battery rooms*, and in 4.3.3 *Electrical work on PV arrays – particular hazards*. Here are some general guidelines:

- refer to 6.6.2 *Electrical work on stand-alone PV systems and particular hazards* below
- all measures regarding battery safety should be strictly observed
- children, unauthorized persons and animals should be kept away from potential hazards
- before carrying out any task, the relevant section of the manufacturer's manual should be read
- components should be checked to make sure they are not damaged
- components should connected/disconnected in the correct sequence
- work should only be done in dry conditions
- no work should be done on PV arrays if there is any lightning in the area
- systems should be inspected and tested before commissioning
- national codes and safe working practices should always be complied with.

6.6.2 Electrical work on stand-alone PV systems and particular hazards

The particular electrical hazards associated with working on PV arrays are discussed in 4.3.3 *Electrical work on PV arrays – particular hazards* and much of what has been said there is relevant to working on stand-alone PV systems. However, PV arrays in stand-alone systems operate at much lower voltages – typically 12 VDC or 24 VDC which reduces the shock hazard (but modules can be connected up incorrectly and give higher voltages). However inverters and inverter-chargers produce mains AC voltages and batteries present additional dangers which are discussed in 6.6.5 *Battery safety.*

Never disconnect the batteries from the charge controller while the PV modules are still connected to the charge controller. The charge controller can be damaged by the high open circuit voltage V_{oc} of the modules. Always disconnect the modules from the charge controller before disconnecting the batteries. During system installation, always connect the batteries to the charge controller before connecting the modules to it.

Inverters are rarely reverse polarity protected. If an inverter is connected up incorrectly to a battery, it can be seriously damaged. Connecting the positive terminal of the inverter to the negative of the battery, and vice versa, can destroy the inverter. This will not be covered by the inverter warranty.

Fuses should not be removed when a circuit is under load – to do so can cause arcing.

Correct DC polarity should be observed when making all electrical connections. Incorrect polarity can damage appliances and system components. All wiring should be inspected and tested before commissioning. This includes testing module/array open circuit voltage V_{oc} and short circuit currents I_{sc}, and verifying DC polarity (see *Testing* in 4.4.1 *DC wiring – modules and strings*) Tests on distribution circuits should be carried out according to national codes.

6.6.3 Lightning protection in stand-alone PV systems

What has been said with regard to lightning protection regarding grid-tied PV systems in 3.7 *Lightning and surge/over-voltage protection* is also largely relevant to stand-alone systems, avoiding cable loops for example. In reality, many small stand-alone PV systems do not have lightning protection. National codes need to be consulted. Modules on roofs or on poles attached to buildings should not be higher than the highest point of the building. The metal module frame above an isolated household can act as a lightning conductor. Charge controllers very often incorporate minimal surge protection such as a voltage suppressor to protect against induced over-voltages. In regions of frequent storms, a double pole disconnect/ isolator should be installed (even on small systems), so that the PV array can be isolated from the rest of the system when there is a risk of lightning strike. No work should be carried out on any PV arrays if there is any thunder or lightning in the area.

6.6.4 Grounding/earthing

It is not usual to ground/earth stand-alone PV systems with one or two modules (< 100 Wp) and with system voltages under 24 VDC, but national codes need to be referred to. Usually the inverter ground/earth terminal (if there is an inverter), the PV array and the negative battery terminal are connected to a single ground/earth electrode. Ground fault circuit interrupters (also known as *residual current devices* or *RCDs)* may or may not be required. For inverter grounding/earthing requirements it is necessary to refer to inverter manuals. The sizes of ground/earthing conductors will usually be specified in national codes. Measures against lightning protection can complicate the requirements.

6.6.5 Battery safety

 Batteries are hazardous. Only suitable trained/qualified persons should work on large battery banks or in battery rooms. The main safety issues are:

- Sulfuric acid is corrosive. In both liquid and vapor form it can cause blindness and damage to skin and mucous membranes (mouth, throat, lungs). Protective goggles and gloves should be worn and a supply of fresh water always at hand – one large bucket or dish for washing and one for rinsing.

- The hydrogen gas produced by vented batteries becomes explosive when mixed with the oxygen in the air. Naked flames, smoking or any sources of sparks can cause ignition. Battery rooms and enclosures should be well ventilated at their highest point (hydrogen wants to rise).(See also 6.6.7 *Battery rooms)*.
- Attention should always be paid to correct polarity when connecting cells / batteries to each other or to inverters and other system components. Short circuits can be made by incorrect connections and reverse polarity can damage equipment, especially inverters.
- While working on batteries, all metal jewelry such as rings, watches, chains, should be removed. These can cause short circuits and melt, causing burns.
- Batteries should never be charged without a charge regulator.
- Tools should be insulated and placed on the floor when not in use. A non-insulated tool falling across the terminals of a battery can cause a short circuit.
- Battery temperatures should not be allowed to exceed 50 °C. They should not be exposed to direct sunlight.
- Used batteries are toxic waste and need to be disposed of correctly. In many countries, battery suppliers are obliged to accept and recycle old batteries.

If the positive pole and the negative pole of a battery cell or battery are connected together with an object made from a conducting material such as a wrench or a screw driver, the battery will short circuit. This can lead to explosion and fire. The utmost care is called for when working on batteries.

6.6.6 Installing and commissioning batteries

Battery manuals and instruction sheets should always be read. These will contain important and exact information on installation, commissioning and maintenance. Lead-acid flooded cells and batteries can be delivered dry or already filled with acid. Batteries with acid in them need to be treated as a hazardous load when they are being transported. Commissioning them is usually straightforward, with only the charge lost through self-discharge since they were originally charged needing to be replaced (but manuals need to be referred to). If batteries are delivered dry, they need to be filled with electrolyte and immediately given an initial commissioning charge. This can involve monitoring SOC and temperature over 3 days. Commissioning batteries in hot climates can be problematic because filling them with acid and ambient temperature means that cell temperatures can exceed 55 °C. When commissioning batteries, cell voltages and electrolyte density should be measured and recorded. (See also *Best practice battery charging* in 6.3.3 *Batteries)*.

Figure 6.16: Batteries should be put in boxes or placed on racks. Battery boxes should have covers to prevent things dropping on them and causing a dangerous short circuit. Battery banks should be in battery rooms (Source: www.hoppecke.com)

6.6.7 Battery rooms

When a system has a very large battery bank (typically consisting of 2 volt cells in large residential or commercial/industrial systems – battery suppliers and manufacturers should be consulted), the batteries should be installed in a separate room, usually adjoining the main building. Battery rooms should be well ventilated, dry, cool yet frost-proof and vibration-free. Ventilation should be natural but in small rooms (or if batteries are installed in a cabinet or cupboard), a ventilation fan may be needed. In fact, in extremely hot or cold climatic conditions, having the battery room below ground level (up to 2 m) can mitigate the excesses of temperature. The floor as well as walls should be protected against the effects of acid – acid resistant plaster or paint can be used. Doors need to open outwards. Door thresholds should be at least 10 cm high. Warning notices should point out the dangers and forbid naked flames/smoking. Batteries should be laid out in such a way that they are easily accessible at least from one side to facilitate the taking of electrolyte readings and maintenance. Battery racks are available from battery manufacturers which hold the batteries securely in place. No equipment such as inverters, controllers, switches or fuses should be installed in the same compartment as the batteries. This equipment may spark and batteries produce corrosive fumes that will destroy expensive hardware. These items should never be installed directly above batteries.

6.7 Stand-alone system maintenance

6.7.1 PV module / array maintenance
Maintaining a PV array on a stand-alone PV system is no different from maintaining an array on a grid-tied system (see 5.3.1 *Maintenance – what system owners can do*).

6.7.2 Battery maintenance
A typical maintenance regime, depending on battery type, at 6-monthly intervals, would include checking that :
- connections are sound and not corroded – petroleum jelly is good protection
- electrolyte level is sufficient – top-up with de-ionized / distilled water if required
- clean top of batteries to remove dirt, dust & moisture
- check cell / battery voltage – in fully charged condition, disconnected from loads and power sources for at least half an hour
- check specific gravity of electrolyte with hydrometer in the event of flooded cells.

All measurements should be recorded. This enables a long-term comparison of the system's performance and enables early recognition of battery aging.

7. Appendices

7.1 Installation codes and guidelines

PV systems need to be compliant with national electrical codes and regulations in their entirety. The documents and parts of documents listed below deal specifically with PV systems, but they need to consulted and interpreted within the general context of the relevant national code or regulations. Neither is this list comprehensive. These documents will also give sources of information on codes covering safe working on building sites, safe working on roofs, lightning protection and other important issues.

USA

Photovoltaic Power Systems and the 2005 National Electrical Code: Suggested Practices, by John Wiles, Southwest Technology Development Institute, copies can be downloaded from http://www.nmsu.edu/~tdi/ Photovoltaics/Codes-Stds/Codes-Stds.html

The 2005 National Electrical Code, Section 720, also *Article 690 Solar Photovoltaic Systems*

Recommended Practice for Utility Interface of Photovoltaic (PV) Systems.

Electrical and Electronic Engineers (IEEE) Standards Board passed IEEE 929-2000

A Guide to Photovoltaic (PV) System Design and Installation, Endecon Engineering and Regional Economic Research for the California Energy Commission, guide to properly installing grid-tied systems has been the model for many Californian installers, PDF 2.4 MB at www.energy.ca.gov/reports/ 2001-09-04_500-01-020.PDF

UK

Photovoltaics in Buildings: Guide to the installation of PV systems, DTI/Pub URN 02/788, can be downloaded at http://www.dti.gov.uk/energy/ renewables/publications/pubs_solar.shtml

Solar Home Systems – developing world

Universal Technical Standard for Solar Home Systems, Thermie B SUP 995-96, EC-DGXVII, 1998, deals only with small stand-alone systems up to 100Wp

7.2 Standards for photovoltaic modules, system components and systems

The following institutions issue standards and photovoltaic modules, system components and systems:

- International Electrotechnical Commission (IEC), www.iec.ch
- Underwriters Laboratories (UL), www.ul.com/info/standard.htm
- American Society for Testing and Materials (ASTM) standards, www.astm.org

7.3 Sizing and design software

PVSOL, *PVexpress*, Valentin Energy Software, www.valentin.de

Solar Dimension, Solaris Energie Consulting, Wolnzach, Germany, www.soldim.de

RETScreen, RETScreen International, www.retscreen.net/

PV design Pro, Maui Solar Energy Software Corporation, www.mauisolarsoftware.com

PV Syst, www.pvsyst.com

(See also *7.6 Sources of information* for some useful websites)

7.4 Site survey form – grid-tied PV system

Sample site survey form for grid-tied PV systems

Customer details

Name ....

Address ....

....

....

....

....

Tel (landline): ....

Tel (cell / mobile): ....

Email ....

Customer requirements

Potential budget: ....

System purpose (exporting to grid mainly? off-setting building electricity consumption? any power back-up needs etc.?)

....

....

....

....

....

Building details
(rough sketch should be made of the building & grounds)

Site longitude : ....

Site latitude: ....

Temperature range at location: ....

Site location (if different from above address): ....

....

....

....

Location of PV array

Free-standing? ...... Is so, distance from building electrical input ....

Roof type ....

Angle .... Orientation .... Height ....

Any shading? ....

Any access problems? ....

Other comments ....

....

....

....

....

....

....

Electrical work details
(supporting diagrams incl. cable length estimates required)

Location of main fuse box, meter, main intake ....

Supply details – VAC, Hz, main fuse, no. of phases etc ....

....

Grounding/earthing system ....

Lightning protection? ....

Possible inverter location ....

Possible location of PV array combiner box ....

Possible location of main DC isolator ....

Utility name & other details ....

Other comments ....

....

....

....

....

....

....

....

....

Survey carried out by ....

Date ....

7.5 Site survey form – stand-alone PV system

Sample site survey form for stand-alone PV systems

Customer details

Name ....

Address ....

....

....

....

....

Tel (landline): ....

Tel (cell/mobile): ....

Email ....

Customer requirements

Potential budget: ....

System purpose (to provide all power? seasonal use? hybrid system? any power back-up needs etc.?)

....

....

....

....

....

Power requirements – lighting and appliances

Light type .... Location .... Power .... (W)

Quantity .... Hours of use per day .... (h)

Light type .... Location .... Power .... (W)

Quantity .... Hours of use per day .... (h)

Light type .... Location .... Power .... (W)

Quantity .... Hours of use per day .... (h)

Light type .... Location .... Power .... (W)

Quantity .... Hours of use per day .... (h)

Light type .... Location .... Power .... (W)

Quantity .... Hours of use per day .... (h)

Appliance .... Location .... Power .... (W)

Quantity .... Hours of use per day .... (h)

Appliance .... Location .... Power .... (W)

Quantity .... Hours of use per day .... (h)

Appliance .... Location .... Power .... (W)

Quantity .... Hours of use per day .... (h)

Appliance .... Location .... Power .... (W)

Quantity .... Hours of use per day .... (h)

Appliance .... Location .... Power .... (W)

Quantity .... Hours of use per day .... (h)

Appliance .... Location .... Power .... (W)

Quantity .... Hours of use per day .... (h)

Water pumping requirements ....

....

Other comments ....

....

....

....

....

Site details (rough sketch should be made of the building(s) & grounds)

Site longtitude: .... Site latitude: ....

Or nearest large city: ....

Temperature range at location: ....

Site location (if different to above address): ....

....

Location of PV array

Free-standing? ..... Is so, distance from building / battery room ....

Or if roof: Type .... Angle .... Orientation .... Height ....

Any shading? ....

Any access problems? ....

Other comments ....

....

Electrical work details
(supporting diagrams incl. cable length estimates required)

Proposed location of batteries ....

Proposed location of charge controller / inverter ....

Distribution wiring ....

Details of any other power sources ....

Other comments ....

....

Water heating requirements
(if any) ....

....

....

Other comments ....

....

Survey carried out by ....

Date ....

7.6 Sources of further information

American Solar Energy Society, www.ases.org

British Photovoltaic Association (PV-UK)

UK PV industry association at www.greenenergy.org.uk/pvuk2

Centre for Alternative Technology (Wales. UK), also books and courses, www.cat.org.uk

Center for Renewable Energy & Sustainable Technology, www.crest.org

Contractors License Reference Site (USA), www.contractors-license.org

Database of State Incentives for Renewable Energy (DSIRE)
Information on USA state, local, utility and selected federal incentives that promote renewable energy, at www.dsireusa.org

The International Solar Energy Society (ISES), www.ises.org

Interstate Renewable Energy Council (USA), www.irecusa.org

Institute for Solar Living (USA), courses, www.solarliving.org

Midwest Renewable Energy Association (USA), courses, www.the-mrea.org

NASA Surface Meteorology and Solar Energy Data Set
Solar radiation data (on horizontal plane) for all locations, at http://eosweb.larc.nasa.gov/sse

North American Board of Certified Energy Practitioners (NABCEP), www.nabcep.org

NREL (National Renewable Energy Laboratory)
Maps of the USA of solar radiation at angle of latitude for each month, at www.nrel.gov and http://rredc.nrel.gov/solar/old_data/nsrdb/redbook/atlas

PVGIS EU Joint Research Centre
Solar radiation data, maps for Europe, Africa, the middle East and Iran, at http://re.jrc.cec.eu.int/pvgis/pv

Solar Energy International (USA), courses, www.solarenergy.org

Solar Energy Industries Assn. (USA), www.seia.org

UK Department of Trade and Industry (DTI) – Renewables
www.dti.gov.uk/renewables

US Department of Energy's Office of Energy Efficiency and Renewable Energy
www.eere.energy.gov/solar

Further reading
Buying a Photovoltaic Solar Electric System, a Consumer Guide, California Energy Commission, PDF 390KB at
www.energy.ca.gov/reports/2003-03-11_500-03-014F.PDF

Got Sun? Go Solar by Rex A. Ewing & Doug Pratt, 2005, concise, richly illustrated up-to-date guide to residential grid-tied solar and wind systems, available from Amazon.com, or direct from the publisher: www.pixyjackpress.com

Home Power, an excellent bi-monthly magazine dealing with solar and wind electrical systems written by practitioners. Current issues can be downloaded from www.homepower.com

Photovoltaics: Design and Installation Manual, Solar Energy International (SEI), for professionals or seriously interested homeowners. Primarily designed as a teaching tool, available directly from SEI: www.solarenergy.org/resources/store.php

Renewable Energy World, free bi-monthly magazine covering all aspects of renewable energy including developments in PV, further information at www.renewable-energy-world.com

Solar Electricity – A Practical Guide to Designing & Installing Small Photovoltaic Systems by Simon Roberts, introduction to stand-alone PV systems, includes tilt correction factor tables for all latitudes and longitudes

Solar Living Sourcebook 12th Edition. Real Goods, 2004. 578 pp, definitive retail guide to renewable energy in North America, part textbook, part catalog.

Small Solar Electric Systems for Africa by Mark Hankins, comprehensive guide to design and installation of SHSs in developing world, stand-alone only

Wind & Sun – Design Guide & Catalogue, part textbook, part catalog, particulary useful for designers and installers of stand-alone PV and PV-hybrid systems, www.windandsun.co.uk

Major PV manufacturers
- BP Solar, www.bpsolar.com
- Evergreen Solar, www.evergreensolar.com/
- GE Solar, www.gepower.com/prod_serv/products/solar/en/index.htm
- Kyocera Solar, www.kyocerasolar.com
- Matrix/Photowatt Solar Technologies, www.matrixsolar.com
- Mitsubishi Electric, http://global.mitsubishielectric.com/bu/solar/index.html
- Photowatt, www. photowatt.com
- Sharp Solar Systems, http://sharp-world.com/solar/
- Shell Solar, http://www.shell.com/home/Framework?siteId=shellsolar
- Uni-Solar, www.uni-solar.com

PV mounting hardware manufacturers
- UniRac, www.unirac.com
- Direct Power and Water, www.directpower.com
- Professional Solar Products, www.prosolar.com/

Inverter manufacturers
- Beacon Power, www.beaconpower.com
- Fronius Solar Electronics,
  http://www.fronius.com/solar.electronics/about/index.htm
- Mastervolt, www.mastervolt.com
- Outback Power Systems, www.outbackpower.com
- PV Powered, www.pvpowered.com
- Sharp Electronics, http://sharp-world.com/solar/
- SMA (Sunny Boy), http://www2.sma.de/index.php?id=20
- Studer, www.studer-inno.com
- Xantrex Technology, www.xantrex.com

Battery manufacturers
- Concorde Battery Corporation, www.concordebattery.com
- Exide, www.exideworld.com
- Hawker Batteries, www.hawkerpowersource.com
- Hoppecke, www.hoppecke.com
- MK Battery, www.mkbattery.com
- Moll, www.moll-batterien.de
- Varta, www.varta

Charge controllers and DC lights
- Morningstar, www.morningstarcorp.com
- Sollatek, www.sollatek.com
- Steca, www.stecasolar.com

Shade analysis aid
- *Solar Pathfinder*, www.solarpathfinder.com

7.7 Power losses and voltage drop in PV systems

Voltage drop is important in PV systems because it can lead to power losses and other problems, particularly in stand-alone systems with lower system voltages. An example of voltage drop would be a situation where a PV module is producing 15 VDC at its output terminals but the voltage reaching the battery terminals is only 13 VDC, which is not enough to charge the battery. Or a situation in which a battery that has a voltage of 13 VDC at its terminals, but the voltage reaching a load is only 11 VDC. This occurs because the cross sectional area of the cables is too small for the current flowing in the cable and the length of the cable. Voltage drop can be calculated as follows:

$$V_D = I \times R_c$$

where

V_D is the voltage drop in the cable (V)

I is the current in the cable (A)

R_c is the cable resistance (Ω), which depends on cable length and cross-sectional area.

EXAMPLE: The voltage at the battery terminals is 12.75 V. A DC television set is drawing a current of 3 A. The total resistance of the cable is 0.4 Ω. What will the voltage drop be and what will the voltage at the appliance terminals be? $V_D = I \times R_c = 0.4\,\Omega \times 3\,A = 1.2\,V$, which is a voltage drop of 9 % (of 12 V). The voltage at the television set would be 12.75 V − 1.2 V = 11.55 V. This also represents a power loss in the cable. $P = V \times I = 1.2\,V \times 3\,A = 3.6$ W, also 9 %. 49 W is being taken from the battery, but only 45.4 W is actually being used by the television set. Power loss in a given cable is proportional to the square of the current flowing in it ($P = I^2 R$). Higher system voltages mean lower currents. This means that cable voltage drop and thus losses in the cables will be less. Doubling the operating voltage means reducing the cross-sectional area of a cable by a factor of 4.

Voltage drop is a problem because:
- in stand-alone PV systems batteries might not be charged properly
- power generated by expensive PV modules is being used to heat up cables instead of doing useful work or going onto the grid
- some appliances may not work properly if the voltage reaching them is too low
- some appliances can be damaged if the voltage reaching them is too low.

Most national electrical codes provide a method of calculating the voltage drop/cable size for the range cables approved in the code by either using a formula or tables. (Moderately inexpensive, but very complete software is available at http://www.softcalcdesigns.com/Products/Efus/Volt_Drop/volt_drop.html). Designers and installers need to be familiar with these methods. Voltage drop also occurs across charge controllers, terminations, switches and fuses, but this is not usually significant.

Different codes use different methods for calculating voltage drop. The methods given here are for common AWG and metric cable sizes. The resistance values given are approximate and are for room temperature (20 °C). Higher ambient and operating temperatures as well as the make of cable can also change the values. First a cable is selected and voltage drop checked. If it is too high the next size cable up is then selected and checked.

Resistances for AWG cable cores

AWG (American Wire Gauge)	Size in mm²	Resistance in ohms per foot (Ω/ft)
14	(2 mm²)	0.002525
12	(3.31 mm²)	0.001588
10	(6.68 mm²)	0.000999
8	(8.37 mm²)	0.000628
6	(13.3 mm²)	0.000395
4	(21.15 mm²)	0.000249
2	(33.62 mm²)	0.000157
1	(42.41 mm²)	0.000127
0	(53.5 mm²)	0.000099

The above table is based on USA copper wire sizes and practice. The following formula can be used: Voltage drop per foot = I × R, where I is the current in the circuit and R is the resistance of the wire. So, in a AWG 12 wire, 300 feet long, carrying a current of 15 A, the voltage drop will be 15 A × 0.001588 Ω = 0.02382 V per foot of cable core. To calculate the voltage drop for a 300 feet cable run, multiply by 600 (feed and return) to get the total voltage drop of 14.29 V.

Resistances for metric size cable cores

Conductor cross-sectional area (mm²)	Resistance in ohms per metre (Ω/m)
2.5	0.0074
4	0.0046
6	0.0031
10	0.0018
16	0.0012
25	0.00073
35	0.00049

The above cable is based on metric copper cable sizes and practice. The following formula can be used: Voltage drop per metre = I × R, where I is the current in the circuit and R is the resistance of the conductor. So, for a 4 mm² cable, 90 m long, carrying a current of 15 A, the voltage drop = 15 A × 0.0046 V = 0.069 V per metre of cable core. To calculate the voltage drop for a 90 metre cable run, multiply by 180 (feed and return) to get the total voltage drop of 12.42 V.

7.8 Abbreviations

α-Si: Amorphous silicon
AC: Alternating current
AM: Air mass
ASTM: American Society for Testing and Materials
BIPV: Building integrated photovoltaics
CHP: Combined heat and power plants
CIS: Copper-indium-diselenide
CdTe: Cadmium-telluride
cm: Centimeter
DC: Direct current

DIY: Do-it-yourself
DOD: Depth of discharge – of a battery
GaAs: Gallium-arsenide
GaSb: Gallium-antimony
GFI: Ground fault interrupter
€:   Euro
h: Hour
HVD: High voltage disconnect – in a charge controller
IEC: International Electrotechnical Commission
I_{sc}: Short circuit current
INV: Inverter
k: Kilograms
km: Kilometer
kW: Kilowatt = 1,000 watts
kWh: Kilowatthour = 1,000 watthour
LED: Light Emitting Diode
LVD: Low voltage disconnect – in a charge controller
m: Meter
mA: Milliamp
mm: Millimeter
MPP: Maximum power point
MPPT: Maximum power point tracking
mS: Millisecond
mV: Millivolt
MW: Megawatt = 10^6 watts = 1,000 kW
PR: Performance Ratio
PV: Photovoltaic
RCCB: Residual current circuit breakers
RCD: Residual current device
SHS: Solar Home System
SOC: State of charge – of a battery
STC: Standard Test Conditions
UL: Underwriters Laboratories
UPS: Uninterruptible power supply
UV: Ultraviolet
V_{oc}: Open circuit voltage
Wp: Peak watt
Wh: Watthour
η: Efficiency
μm: Micrometer

(See also 7.9 *Physical units*.)

7.9 Physical units

Property	Unit
Current I	A: Ampere
Voltage V	V: Volt
Resistance R	Ω: Ohm
Impedance Z	Ω: Ohm
Power P	W: Watt kW: Kilowatt = 1,000 W = 10^3 W MW: Megawatt = 1,000,000 W = 1,000 kW = 10^6 W GW: Gigawatt = 1,000,000,000 W = 1,000,000 kW = 10^9 W
Energy	Wh: Watthour = 0.001 kWh kWh: Kilowatthour = 1,000 Wh MWh: Megawatthour = 1,000,000 Wh = 1,000 kWh GWh: Gigawatthour = 1,000,000,000 Wh = 1,000,000 kWh TWh: Terawatthour = 1,000,000,000,000 Wh = 1,000,000,000 kWh
Frequency f	Hz: Hertz
Solar power (intensity)	W/m²: Watts per square meter
Solar energy / peak sun hour	kWh/m²: Kilowatthours per square meter
Solar energy/annual	kWh/m²/year: Kilowatthours per square meter per annum

7.10 Index